STUDY GUIDE AND COLOR

David S. Smith

FUNDAMENTALS OF ANATOMY AND PHYSIOLOGY

Frederic Martini

PRENTICE HALL, ENGLEWOOD CLIFFS, NEW JERSEY 07632

Editorial/production supervision: Shelia Whiting
Manufacturing buyer: Paula Massenaro

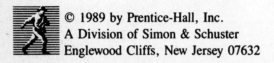
Printed in the United States of America

10 9 8 7 6 5 4 3 2 1

ISBN   0-13-444415-9

Prentice-Hall International (UK) Limited, *London*
Prentice-Hall of Australia Pty. Limited, *Sydney*
Prentice-Hall Canada Inc., *Toronto*
Prentice-Hall Hispanoamericana, S.A., *Mexico*
Prentice-Hall of India Private Limited, *New Delhi*
Prentice-Hall of Japan, Inc., *Tokyo*
Simon & Schuster Asia Pte. Ltd., *Singapore*
Editora Prentice-Hall do Brasil, Ltda., *Rio de Janeiro*

Contents

TO THE STUDENT

Welcome to the world of human anatomy and physiology! You are about to begin a study one of the most fascinating subjects known, the structure and function of the human body. You are fortunate in that you will be using one of the finest resources available, Fredrick Martini's textbook of human anatomy and physiology. This study guide and workbook is designed to help you master the material contained in the textbook and in doing so, master the subject of human anatomy and physiology.

The subject of human anatomy and physiology is interesting, but it is also complex. It is a subject that requires a great deal of effort on your part. There are many facts and principles that must be mastered and the function of this study guide and workbook are to facilitate the mastery of the subject.

Organization of the Workbook and Study Guide

The organization of this study guide parallels that of the textbook. There are twenty-nine chapters, one for each of the textbook chapters. Each chapter of the study guide is divided into four parts.

The first part is entitled "overview." This consists of one to several paragraphs in which the material in the chapter under study is briefly described. In addition to the chapter content, the relevance of the material and how it fits in with other topics is also discussed.

The second part is a detailed outline of the chapter. This includes all of the major headings and subheadings included in the chapter. This provides a more detailed look at the chapter contents plus an organizational framework of the material.

The third part is entitled "learning activities." This is the heart of the study guide as it is the section that will lead you through the chapter. The learning activities consist of a number of items for you to complete. These include fill in the blank, matching, diagram labeling, and coloring exercises. The sequence of the items matches their appearance in the textbook chapter. Therefore, by working through a chapter in the study guide, you will in essence be working through the textbook chapter in the same order. Although not every fact or principle presented in the textbook will appear in the study guide, you will find that the bulk of the basic anatomy and physiology will be covered.

The labeling and coloring items deserve special comment. These are for the most part anatomically oriented exercised. Usually a diagram will be presented with blank labels which you are to complete. Most of these illustrations will also require coloring. There are several reasons for these coloring exercises. First, they are relaxing. Almost all of us enjoy coloring pictures. This in itself provides a "change of pace" from the grind of learning difficult material. There is another reason besides fun and games. When you color in an anatomical structure you are forced to pay very close attention to all of it. You focus your attention on aspects of it that you would miss in a simple labeling exercise. Consequently you will learn a lot of anatomy without even being aware that you are learning it! You can make these exercises even more beneficial by repeating to yourself the name of the structure and its function while your are coloring it. The best coloring instruments are colored pencils. A box of about 8 colors is all you need.

The fourth and final part of each chapter is the self test. This consists of fifteen multiple choice questions about the material contained in the chapter. You should be able to answer twelve of the fifteen (80%) without referring to the text or study guide chapter. If you find that you cannot answer that many then you probably need to do some additional work on that chapter.

Study Hints

Studying any subject, especially one as complex as human anatomy and physiology, is not a hit or miss proposition. There are systematic habits that can be developed which will greatly increase the efficiency of the studying process. Some of these are listed below.

1. Make sure that you choose a suitable place for study. A comfortable desk or chair with good, over the shoulder lighting is ideal. When working with the study guide you will be writing and coloring. Consequently a flat surface is useful to have. A desk or table top of comfortable height will do nicely. There should not be a lot of distractions in your study area. Loud talking, televisions, and blaring radios detract from the learning process.

2. Establish a regular schedule of study. You should set aside thirty to sixty minutes each day for study. This is far better than trying to study for three hours at a stretch on one or two days of the week. As you will learn during your studies, neural synapses fatigue when used continuously for extended periods of time. Once that happens you are accomplishing nothing in terms of learning.

3. Start each study session by quickly reviewing the previously covered material. The key to mastery is repetition. A quick review within 24 hours of having studied new material will greatly reinforce what was learned. A very easy way to do this is to read through the study outline that is provided at the end of each chapter of your textbook.

4. Develop a systematic approach to studying. Science is a systematic way of looking at the world, and your approach to learning science should also be systematic. The approach detailed below is one that has proven very useful to a large number of students.

a. Begin each chapter by first reading the questions posed by the author at the begining of the chapter. These will tell you what the author considers to be the important points.

b. Read the overview and chapter outline in the study guide. This will give you a good idea of exactly what is contained in the chapter. Do not be upset if you do not understand all of the terms or expressions used in the outline. You will learn these as you read the chapter. The very fact that you know they are coming will give them more meaning and impact when you do read about them.

c. Read the chapter in the textbook. As your read the chapter pay particular attention to the terminology. One of the things that makes anatomy and physiology difficult for the new student is the extensive vocabulary need to describe its aspects. You will be introduced to more new terms in this course than are found in the typical foreign language course. They must be mastered and you should begin immediately. Also make sure that you pay attention to the illustrations and tables. A large amount of information is contained in these, and to bypass them eliminates about fifty percent of the information in the textbook. After you have completed reading the text, be sure to read the study outline at the chapter's end. This study outline is a detailed summary of all of the major topics included in the chapter. It constitutes an excellent review, and is one of the outstanding features of your textbook.

d. Once you have completed reading the chapter, proceed to the learning activities in the study guide. If you find an item that you are not sure about try to find it in the textbook first. The answers in the back of the book are "last resort." You are much better off spending the time to review the textbook and find the answer there. You will find that many of the learning exercises will refer you to a figure or table in the textbook for assistance. This is always true for the illustrations. If you experience difficulty with these then by all means go directly to the referenced figure or table for help. Some of these exercises, especially the ones referencing tables, are designed to force you to read carefully a particular table which contains important information.

e. Upon completion of the of the learning activities you may want to refer to the questions contained at end of the chapter in your textbook. They constitute another source of review and assistance.

f. Complete the self test in the study guide. Once you have answered all fifteen questions, check your answers against the answer key at the end of the study guide. The self test answer key is the last part of the answer section for each chapter. If you have mastered the material, you should be able to answer twelve of the fifteen questions. Be sure that you look up in your textbook any questions that you miss.

g. If you are studying this material as part of a formal college course then you will in all probability be attending formal lectures. By all means be sure that you have read the chapter before attending lecture. You will find that the lecture is much more meaningful to you if you have read the material that is to be discussed. Another technique that you may find useful is to copy down the chapter outline from this study guide into your lecture notebook. Leave space beneath each heading and subheading, and then fill in with your own notes as the various subjects are discussed. This will not only organize your note taking, but it will save you much writing time during lecture, time then can be much better spent carefully listening and thinking about the subjects being discussed.

Finally, do not become discouraged. To the neophyte student the plethora of mysterious terms and polysyllable words that characterizes human anatomy and physiology is indeed intimidating. Keep in mind that with a systematic approach to study and sufficient time, any average person can master the subject. The secret is to stay with it. As you proceed through the subject you will find that mastery of previous material makes the new material much easier, and eventually you will develop a "critical mass" of information which will permit you to proceed with very little impairment. The rewards of success are worth the time and energy. Besides all of the practical applications, there is a great deal of satisfaction in understanding how you, yourself, are built and how you operate.

CHAPTER 1
An Introduction
to Anatomy and Physiology

OVERVIEW

This chapter places the study of anatomy and physiology into its scientific context. The basic elements of anatomy and physiology are identified and the major themes that will guide your study—form and function, levels of organization, and homeostasis—are introduced. A brief historical narrative traces the development of anatomy and physiology from its origin in Ancient Greece to the present. The chapter concludes with an essay on the nature of science that you should examine carefully. This essay establishes the basic principles that guide science and you should master these principles and apply them to all of your future studies and professional activities.

CHAPTER OUTLINE

    A. Introduction
    B. The science of anatomy and physiology
        1. Topics in anatomy and physiology
        2. Themes and patterns in anatomy and physiology
            a. Structure and function
            b. Levels of organization
            c. Homeostatic regulation
            d. Homeostasis and disease
            e. Homeostasis and development
    C. Historical perspectives in anatomy and physiology
    D. Essay - The scientific method

LEARNING ACTIVITIES

Complete the following items by supplying the appropriate word or phrase.

1. All living things exhibit seven basic characteristics. These characteristics are

   _____

   _____

   _____

   _____

   _____

   _____

   _____

2. Biology is defined as the study of _____.

3. The branch of zoology that is concerned with the internal and external structure of animals is _____.

4. Physiology is the study of _____.

5. Match the correct term from the list below with the statements that follow.

   a. gross anatomy   b. microscopic anatomy   c. surface anatomy   d. systemic anatomy   e. regional anatomy   f. medical anatomy   g. radiographic anatomy   h. pathological anatomy   i. surgical anatomy

   ____ the study of anatomical landmarks important for surgery

   ____ anatomy as determined by penetrating radiations

   ____ the study of abnormal structures

   ____ anatomy that can be studied with the naked eye

   ____ the study of the structure of the major organ systems

   ____ anatomy that requires the aid of a microscope

   ____ the study of general form and superficial markings

   ____ the examination of all of the structures of a body region

6. The study of cells best describes _____.

7. The study of tissues is referred to as _____ while the study of whole organs is known as _____.

8. The study of anatomy and physiology together is extremely useful because in biology _____ and _____ are always closely related.

9. Arrange the following levels of organization into their proper sequence by placing a number in front of each one. The most basic level should be numbered one while the most complex will be numbered seven.

   ____ tissue

   ____ cell

   ____ molecule

   ____ atom

   ____ organ

   ____ organism

   ____ organ system

10. The maintenance of a constant and stable internal environment within the body is known as _____.

11. The component of a homeostatic regulatory system that is sensitive to changes in the environment (stimuli) is the _____.

12. The _____ of a homeostatic regulatory system acts upon the stimulus for that system.

13. In the control system illustrated below, label the receptor, effector, and stimulus.

    temperature   thermostat   heater

    a. _____, b. _____ c. _____

14. In the example above, the heater produces heat which increases the temperature and ultimately causes the thermostat to shut the heater off. This is an example of _____ feedback.

15. The increasingly powerful muscle contraction associated with childbirth labor is an example of _____ feedback.

16. Failures in homeostasis result in _____.

17. It is important for health care personnel to understand normal homeostatic mechanisms because that will permit them to predict the key features of potential _____.

18. The programmed structural and functional changes that occur within our bodies over time are part of the process of _____.

19. The study of developmental processes during the first two to three months of development is known as _____.

20. Developmental errors that affect homeostatic mechanisms sufficiently to cause clinical symptoms at birth are termed _____ _____.

21. Match the person listed below with the contributions to anatomy and physiology that follow.

    a. Galen b. Harvey c. Hippocrates d. Rhazes e. Vesalius

    ____ wrote volumes on anatomical topics around 400 B.C.

    ____ famous Roman physician and anatomist who wrote books that were used in medical studies through the Dark Ages

    ____ discovered the circulation of blood

    ____ Arab physician who added to Galen's work

    ____ sixteenth century anatomist who made highly detailed anatomical drawings

22. A tentative explanation of a group of data or facts best defines _____.

23. The three characteristics that a valid hypothesis must have are _____, _____, and _____.

24. A hypothesis that has met all three criteria for validity will be accepted as a scientific _____.

25. Scientific theories make accurate _____ about the real world.

SELF TEST

Circle the correct answer to each question.

1. Which of the following is not a common characteristic of all living things?
   a. responsiveness b. growth c. absorption d. reproduction e. circulation

2. Anatomy that can be studied with the naked eye is termed
   a. microscopic. b. regional. c. radiographic. d. pathological. e. gross.

3. A medical student is assigned to learn all of the named structures found in the arm of a cadaver. The type of anatomy that he will be studying is known as _____ anatomy.

   a. gross   b. microscopic   c. regional   d. radiographic   e. pathological

4. The study of tissues best describes the science of

   a. cytology.   b. histology.   c. organology.   d. biology.   e. physiology.

5. An anatomist will examine the structure of a body part. The physiologist will be concerned with its

   a. shape.   b. size.   c. length.   d. function.   e. weight.

6. The part of a homeostatic control system that responds to a stimulus is the

   a. receptor.   b. effector.   c. positive feedback.   d. negative feedback.   e. none of the above

7. In a negative feedback control system, the effector

   a. magnifies the stimulus.   b. increases the stimulus intensity.   c. decreases or eliminates the stimulus.   d. has no effect on the stimulus.   e. shuts down the receptor.

8. Loss of homeostasis usually results in

   a. little change in the body.   b. a decrease in receptor sensitivity.   c. an increase in negative feedback.   d. disease.   e. none of the above

9. Knowledge of homeostatic mechanisms makes it possible to

   a. predict disease symptoms.   b. identify physiological receptors.   c. identify physiological effectors.   d. understand the origin of disease processes.   e. more than one of the above is correct

10. The events that occur during the first two or three months of development are the subject matter of

    a. histology.   b. embryology.   c. cytology.   d. organology.   e. physiology.

11. The Greek who wrote on medical and anatomical topics around 400 B.C. was

    a. Galen.   b. Harvey.   c. Vesalius.   d. Hippocrates.   e. Avicenna.

12. The man who discovered the circulation of blood was

    a. Galen.   b. Harvey.   c. Vesalius.   d. Hippocrates.   e. Rhazes.

13. The man responsible for the first modern anatomical illustrations was

    a. Galen.   b. Harvey.   c. Vesalius.   d. Hippocrates.   e. Rhazes.

14. A tentative explanation of a number of facts best defines a (an)

    a. fact.   b. principle.   c. hypothesis.   d. theory.   e. idea.

15. A hypothesis that is unbiased and has been repeatedly tested and found to be true will usually be accorded the status of a (an)

    a. fact.   b. scientific theory.   c. scientific law.   d. general principle.   e. confirmed hypothesis.

CHAPTER 2
The Chemical Level
of Organization

OVERVIEW

In this chapter the most basic level or organization that is found in the living world is examined. This level consists of the atoms and molecules, the substances which comprise the chemical level of organization. All of life's processes can be ultimately explained at this level, and it is at this level that all disease processes begin. A thorough knowledge of this level is absolutely essential for an understanding of all of the remaining levels which comprise the body. It is assumed that you have little background in chemistry: therefore this chapter begins with the most fundamental aspects of matter, proceeds through the basics, and finally into the more complex world of the organic molecules which constitute the substance of living tissues. It is important that you follow the material in sequence through this chapter. This is because each chemical concept that is introduced depends upon the one that preceded it for understanding. You may find this chapter difficult, but it contains material that absolutely must be mastered if you are to comprehend all that will follow.

CHAPTER OUTLINE

    A. Introduction
    B. Atoms and molecules
    C. Atomic interactions
       1. Covalent bonds
          a. Polar bonds
       2. Ionic bonds
       3. Hydrogen bonds

   D. Chemical notation
   E. Chemical reactions
   F. Chemical organization of the human body
        1. Inorganic compounds
             a. Water and minerals
             b. Electrolytes and solutions
             c. pH of biological solutions
             d. Strong acids and bases
   G. Organic compounds
        1. Carbohydrates
        2. Lipids
        3. Proteins
             a. Protein structure
             b. Protein function
             c. Enzyme structure and function
             d. Enzymes and homeostasis
             e. Other special proteins
        4. Nucleic acids
        5. Miscellaneous organic compounds
             a. High energy compounds
             b. Vitamins
             c. Porphyrins

LEARNING ACTIVITIES

   Complete the following items by supplying the appropriate word or phrase.

   1. Matter is anything that occupies space and has _____.
   2. The smallest stable unit of matter is the _____.
   3. Atoms in turn are composed of three fundamental particles, the _____, _____, and _____.
   4. Match the correct term from the list below with the statements that follow.

       a. proton b. neutron c. electron d. atomic number e. element f. atomic weight g. energy levels h. nucleus

       _____ possesses a negative charge

       _____ the total number of protons and neutrons in the atom

       _____ possesses a positive charge

       _____ subatomic particle that has no electrical charge

       _____ the center of the atom

       _____ the regions where electrons are found

       _____ the number of protons found in an atom

       _____ atoms that have the same atomic number

   5. The first energy level or shell can hold _____ electrons and the second contains a maximum of _____.
   6. The atomic weight of sodium is 23. One mole of sodium would contain _____ grams.
   7. Isotopes are atoms of an element that contain different numbers of _____.

8. _____ have unstable nuclei that emit subatomic radiation in measurable amounts.

9. The chemical properties of an element are determined by its _____ electron shell.

10. Atoms can interact to form molecules by _____, _____, or _____ electrons in their outer shell.

11. The overall result of the formation of a chemical bond is the filling of the _____ shells of the participating atoms.

12. A _____ is a molecule made up of two or more different kinds of atoms.

13. Atoms share pairs of electrons to form _____ bonds.

14. Two atoms sharing two pairs of electrons would create a _____ covalent bond.

15. Covalent bonds in which the pairs of electrons are shared unequally produce _____ bonds that result in the poles of the molecule having electrical charges.

16. The transfer of an electron from one atom to another forms an _____ bond.

17. The loss or gain of electrons by an atom results in the production of _____.

18. Ions bearing a positive charge are termed _____, while those bearing negative charges are termed _____.

19. Ionic compounds are held together by the _____ attraction of the ions involved.

20. The number of charges associated with an ion is termed its _____.

21. An atom that contained seven electrons in its outer shell would most likely form a _____ ion.

22. Hydrogen bonds form between hydrogen and _____ or _____.

23. Hydrogen bonds are important in that they determine the _____ of complex organic molecules.

24. The atomic weight of hydrogen is 1 and that of oxygen is 16. The molecular weight of water (HOH) would be _____.

25. For the following reactions write the category of the reaction in the space provided.

    _____ C + D = CD
    _____ CD = D + C
    _____ CD + EF = CE + FD
    _____ CDEF = CD + EF
    _____ C + D + F = CDF

26. Chemical reactions that release energy are _____ while those that absorb energy are _____.

27. Inorganic compounds do not contain _____ and _____ as their primary structural elements.

28. Unequal sharing of the electrons between hydrogen and oxygen atoms causes water molecules to be _____.

29. _____ bonding gives water a remarkable ability to absorb and retain heat.

30. In a solution, the dissolved compound is termed the _____.

31. When ionic compounds and some polar covalent compounds are dissolved in water they _____ to form ions.

32. Another term for ions is _____ because they will conduct an electric current.

33. The _____ of a solution is equal to the negative exponent of its hydrogen ion concentration.

34. A solution with a pH of 2 would have more _____ than a solution with a pH of 3.

35. A pH value of _____ is considered neutral.

36. Adding an acid to a solution always _____ its pH value.

37. A strong acid or base is one which _____ completely.

38. A compound which liberates a hydroxyl ion in solution is termed a _____.

39. Acids react with bases to form _____.

40. _____ maintains pH homeostasis by removing or replacing hydrogen ions.

41. Organic compounds always contain _____ and _____ atoms.

42. The sum total of all reactions that occur in the organism defines _____.

43. _____ is that part of metabolism that consists of decomposition reactions which provide the _____ necessary for life.

44. Synthetic reactions make up that branch of metabolism known as _____.

45. Carbon atoms can form long chains by forming _____ with each other.

46. Each carbon atom can form _____ covalent bonds.

47. Organic compounds that will not dissolve in water are _____ while those which dissolve in water are _____.

48. A carbohydrate is composed of carbon, hydrogen, and oxygen in a ratio near _____.

49. For the monosaccharides listed below, write the number of carbons found in each one in the space provided.

    _____ triose
    _____ tetrose
    _____ pentose
    _____ hexose
    _____ heptose

50. The hexose _____ is the most important fuel molecule in the body.

51. Sucrose is made up of two monosaccharides and is an example of a _____.

52. Two molecules are bonded together and in the process a molecule of water is generated. This is an example of _____ synthesis.

53. The molecule formed in question 52 could be broken down into its original components by adding _____, a reaction process known as _____.

54. _____ are large complex molecules which are made up of many smaller carbohydrate units.

55. An example of a polysaccharide found in our bodies would be _____.

56. Fats, oils, and waxes belong to the class of organic molecules known as the _____.

57. The organic acid group (COOH) which gives fatty acids their acidic properties is termed a _____ group.

58. A fatty acid which contained two carbon atoms that shared two covalent bonds between them would be described as _____.

59. _____ acids are those that cannot be synthesized by our body.

60. Glycerol + 2 fatty acids equals a _____.

61. Another name for triglycerides is _____.

62. Fats function as long term _____ storage compounds.

63. _____ are lipids that function as "local hormones."

64. Steroid hormones are derived from the compound _____.

65. Cholesterol is a major component of cell _____.

66. The three major structural lipids are _____, _____, and _____.

67. _____ are the basic building blocks of proteins.

68. There are _____ different amino acids.

69. The bond formed between amino acids by dehydration synthesis is known as the _____ bond.

70. Three amino acids bonded together would form a _____.

71. _____ proteins are insoluble in water and play a structural role in the body.

72. _____ proteins are compact, rounded, and soluble in water.

73. Match the terms below with the appropriate statement that follows.

a. primary structure   b. secondary structure   c. tertiary structure   d. quaternary structure

_____ often results in an alpha-helix structure

_____ the sequence of amino acids in a protein

_____ interactions between polypeptide chains

_____ provides the three-dimensional shape of a protein

_____ determines the functional properties of a protein

_____ is destroyed during denaturization

74. In the spaces provided below, list the major functions of proteins in the body.

_____

_____

_____

_____

_____

_____

_____

75. _____ _____ is the amount of energy that must be put into all chemical reactions before they will occur.

76. A _____ speeds up a chemical reaction but is not affected by that reaction.

77. _____ function as biological catalysts.

78. In an enzyme-catalyzed reaction the reactants are termed _____.

79. The tertiary structure of an enzyme creates an _____ in which the substrate binds.

80. Most enzyme-catalyzed reactions require _____ in addition to the enzyme and substrate.

81. _____ are complex organic cofactors which are frequently derived from vitamins.

82. The major significance of homeostasis is the maintenance of a constant environment to insure proper _____ functioning.

83. _____ are combinations of large protein and small carbohydrates while _____ are made up of large carbohydrates and short polypeptide chains.

84. The two major types of nucleic acids are _____ and _____.

85. The three components of a nucleotide are a _____, a nitrogen base, and a _____ group.

86. The sugar of RNA is _____ and that of DNA is _____.

87. The four nitrogen bases of DNA are _____, _____, _____, and _____.

88. The four nitrogen bases of RNA are _____, _____, _____, and _____.

89. In the synthesis of a nucleic acid, dehydration synthesis attaches the _____ group one nucleotide to the _____ of another.

90. RNA is single stranded and DNA is _____ stranded.

91. A strand of DNA has a nucleotide sequence of A-T-C-G. Its complementary strand would have the sequence _____.

92. The principal high energy compound produced during catabolism is _____.

93. The conversion of _____ to ATP represents the primary method of energy storage in our cells.

94. Because _____ cannot be synthesized by our bodies, they must be in our diets.

95. _____ is the porphyrin that gives blood its color.

SELF TEST

Circle the correct answer to each question.

1. Carbon has four electrons in its outer shell. It can form _____ covalent bonds.
   a. 1 b. 2 c. 3 d. 4 e. 5

2. A chemical bond that is formed by the transfer of an electron from one atom to another is a (an) _____ bond.
   a. covalent b. ionic c. hydrogen d. strong e. coordinate

3. Hydrogen bonds are weak electrical attractions between hydrogen and
   a. chlorine or fluorine. b. carbon or silicon. c. carbon or chlorine. d. oxygen or nitrogen. e. carbon or phosphorus.

4. The reaction AB + CD = AC + BD is an example of a (an) _____ reaction.
   a. decomposition b. synthesis c. exchange d. exergonic e. endergonic

5. The bulk of our body weight is due to
    a. proteins. b. lipids. c. carbohydrates. d. water. e. minerals.

6. Which of the following pH values represents the greatest amount of hydrogen ion?
    a. 7 b. 5 c. 4 d. 3 e. 2

7. Compounds that dissociate in water form
    a. amino acids. b. ions. c. sugars. d. proteins. e. nucleotides.

8. Organic compounds always contain
    a. carbon-hydrogen. b. carbon-nitrogen. c. carbon-oxygen. d. carbon-phosphorus. e. none of the above

9. The class of organic compounds which are the most important source of cellular energy are the
    a. carbohydrates. b. lipids. c. proteins. d. nucleic acids. e. porphyrins.

10. Two simple sugars can form dehydration bonds with each other and yield a
     a. monosaccharide. b. disaccharide. c. polysaccharide. d. dipeptide. e. triglycerol.

11. The compound that gives rise to the steroid hormones is
     a. triglycerol. b. fatty acid. c. prostaglandin. d. cholesterol. e. phospholipid.

12. The three-dimensional shape of a protein results directly from the
     a. primary structure. b. secondary structure. c. tertiary structure. d. quaternary structure. e. none of the above

13. Enzymes catalyze biochemical reactions by
     a. increasing the energy of activation. b. decreasing the energy of activation. c. increasing the free energy of the reaction. d. decreasing the free energy of the reaction. e. forming covalent bonds with the substrate.

14. The region of the enzyme that interacts with the substrate is the
     a. coenzyme. b. active site. c. primary sequence. d. polypeptide subchain. e. secondary structure.

15. The complementary sequence of the DNA strand C-C-G-G would be
     a. C-C-G-G. b. C-C-C-C. c. G-G-C-C. d. G-G-T-A. e. T-T-A-A.

CHAPTER 3
The Cellular Level
of Organization

OVERVIEW

One of the broadest generalizations in biology is the cell theory. This theory states that all living things are composed of cells and/or cell products, and that cells are derived from preexisting cells. In other words, the cell is the structural, functional, and reproductive unit of life. Our bodies are composed of cells and it is within these cells that the chemical reactions of life occur. Ultimately it is the individual cell which reproduces itself and our multicellular bodies represent many millions of these replications. The cell is also the first level of complexity which is considered alive. Another way of stating this is that the cell represents the minimal amount of chemical organization that can execute all of the basic functions of life. Obviously a good understanding of this building block of life is essential if you are to comprehend the structure and function of the body.

This chapter introduces you to the major aspects of cell biology. It begins by explaining the structure of cells, how they are put together. Once you have mastered cellular structure you will then move onward to cell functioning. First you will examine how materials move into and out of the cell. Then you will utilize the material which you learned in Chapter 2 and examine the major chemical reactions that go on in the cell, both the energy yielding reactions and the energy consuming reactions. These are the fundamental reactions of life. Finally you will survey how cells reproduce themselves.

CHAPTER OUTLINE

A. Introduction
    1. Cytological techniques
B. Cellular anatomy
    1. The cell membrane

2. The cytoplasm
    a. The cytosol and inclusions
    b. Organelles
        (1) The cytoskeleton
        (2) Centrioles, cilia, and flagella
        (3) The mitochondria
        (3) The nucleus
        (4) Ribosomes
        (5) The endoplasmic reticulum
        (6) The Golgi apparatus
        (7) Lysosomes and peroxisomes
    c. Membrane interactions
B. Cellular physiology
    1. Passive membrane processes
        a. Diffusion
        b. Osmosis
        c. Filtration
    2. Active membrane processes
        a. Active transport
        b. Endocytosis
    3. The transmembrane potential
C. Cellular metabolism
    1. Catabolic processes
        a. Phosphorylation
            (1) Substrate-linked phosphorylation
            (2) Oxidative phosphorylation
        b. Mitochondria and energy production
        c. Carbohydrate catabolism
            (1) Aerobic glycolysis
            (2) Anaerobic glycolysis
        d. Lipid catabolism
        e. Protein catabolism
        f. Nucleic acid catabolism
    2. Anabolic processes
        a. Carbohydrate synthesis
        b. Lipid synthesis
        c. Protein synthesis
            (1) The nucleus and transcription
            (2) Translation
        d. Nucleic acid synthesis
    3. The regulation of cellular metabolism
D. Cellular reproduction and diversity
    1. Mitosis
    2. The origin of cellular diversity
E. Essay: Gene Identification and Manipulation

LEARNING ACTIVITIES

Complete the following items by supplying the appropriate word or phrase.

1. The study of cellular structure and function is known as _____.

2. The maximum magnification of the typical light microscope is about _____ times.

3. In _____ electron microscopy, electrons pass through an ultrathin section, while _____ electron microscopy bounces electrons off exposed surfaces.

4. Label Figure 3.1 using the structures listed above it. Color each of the structures, using a different color for each one. Refer to Figure 3.3 in your textbook for assistance.

Cell membrane	Cytosol	Nucleus
Mitochondria	Ribosomes	Endoplasmic reticulum
Golgi apparatus	Lysosomes	Microvilli
Cilia		

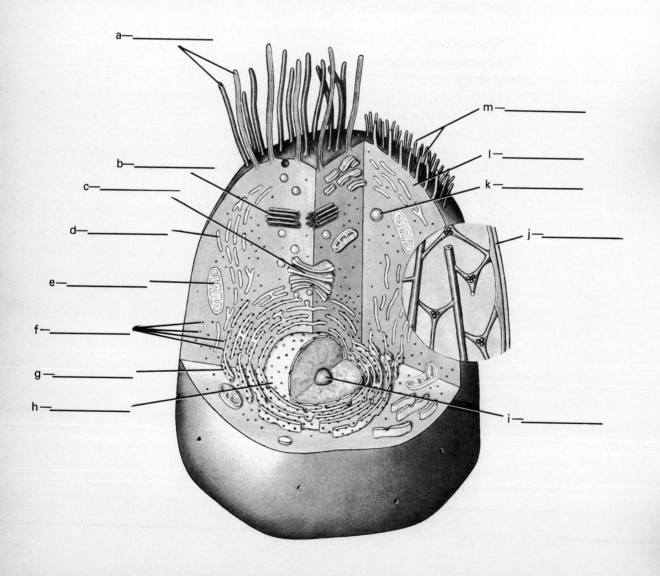

Figure 3.1

5. Complete the following table of cellular functions with the terms which were used in labeling 3.1. Refer to Table 3.1 of your text for assistance.

_____ moves material over surfaces

_____ isolates, protects, and makes the cell sensitive to its environment

_____ controls metabolism: stores and processes genetic information

_____ distributes material by diffusion

_____ membrane sacs containing digestive enzymes

_____ produces the bulk of the cell's ATP

_____ the site of ribosome synthesis

_____ moves the chromosomes during mitosis

_____ function as the sites of protein synthesis

_____ packages cell secretions

_____ functions in intracellular transport and synthesis

_____ absorption of extracellular material

6. The hydrophobic ends of the lipid molecules found in the cell membrane are oriented towards the _____.

7. _____ proteins are parts of the membrane and extend into the hydrophobic layer.

8. _____ guarded by integral proteins allow water and specific electrolytes to travel in or out.

9. The _____ and _____ form a cytoskeleton and also function in the movement of cellular structures.

10. The cilia originate from the _____ which is located in the peripheral cytoplasm.

11. Microtubules are composed of globular proteins termed _____.

12. _____ consists of 9 groups of microtubules arranged in a circle.

13. The inner membrane of the mitochondrion is folded to form _____.

14. The movement of materials from areas of high concentration to areas of low concentration is known as _____.

15. The four major factors that determine whether a molecule will diffuse across a cell membrane are _____, _____, _____, and _____.

16. Facilitated diffusion differs from ordinary diffusion in that a _____ molecule is required.

17. The major ion involved in powering cotransport systems is _____.

18. _____ is the movement of water across membranes in response to differences in the concentration of water molecules.

19. Match the following statements with the terms hypotonic, isotonic, and hypertonic.

_____ cells will shrink in this solution

_____ cells will swell and burst in this solution

_____ cells will remain unchanged in this solution

_____ a solution with a greater concentration of water than a cell

_____ a solution with a lesser concentration of water than a cell

_____ injections are usually given with this kind of solution

20. _____ is the forcing of a solution across a membrane by hydrostatic pressure.

21. Movement of a molecule against a concentration gradient is known as _____ and requires the expenditure of _____.

22. Active transport requires specific _____, _____ molecules, and _____.

23. Ingestion of fluid droplets is known as _____ while ingestion of solid matter is termed _____.

24. The cell membrane normally has an excess of _____ ions on the outside of the membrane and an excess of _____ ions on the inside of the membrane.

25. The separation of negative and positive charges across the membrane creates a _____.

26. The difference of electrical potential that exists across all cell membranes is termed the _____ potential.

27. Changing of the transmembrane potential is the trigger for initiating the _____ of muscle.

28. The study of how cells obtain and use energy is the study of _____.

29. Metabolic turnover, growth, and _____ are examples of anabolism.

30. _____ is the breakdown of large molecules into smaller units with a release of energy.

31. Catabolic reactions provide energy for _____ in which inorganic phosphate is attached via a high energy bond to ADP to create ATP.

32. In addition to ATP there are four other high energy compounds created by phosphorylation reactions. They are _____, _____, _____, and _____.

33. In _____ phosphorylation, an enzyme breaks a molecular bond, and the energy released attaches a molecule of inorganic phosphate to the substrate.

34. Oxidative phosphorylation begins with a pair of hydrogens which are removed from a _____ molecule.

35. Coenzymes involved with the removal and transport of hydrogen include _____, _____, _____, and _____.

36. During oxidative phosphorylation, the hydrogen which is removed ionizes and donates an electron to a special complex called the _____.

37. The participants in the electron transport system are metaloproteins called _____.

38. Every time electrons move from one member of the electron transport system to another, _____ is lost.

39. The energy lost during the transport of electrons can be used to synthesize _____.

40. The final electron acceptor in the respiratory chain is _____.

41. The final compound produced by the electron transport system is _____.

42. Oxidative phosphorylation is the single most important mechanism for the generation of _____.

43. Complete Figure 3.2, oxidative phosphorylation, by supplying the missing information. Refer to Figure 3.20 in your textbook for assistance.

44. The major source of hydrogen ions for oxidative phosphorylation are the chemical reactions that occur within the _____.

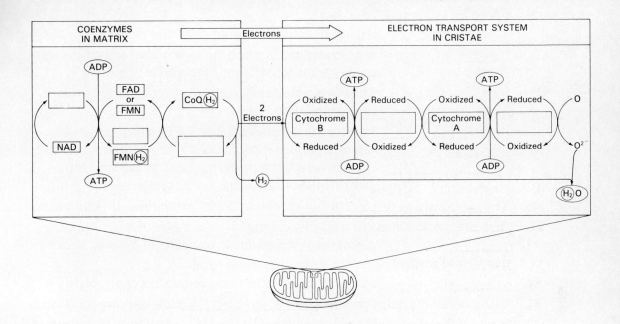

Figure 3.2

45. The enzymes of the _____ are found in the mitochondrial matrix.

46. Complete Figure 3.3, the Kreb's cycle, by supplying the missing information. Refer to Figure 3.21 of your text for assistance.

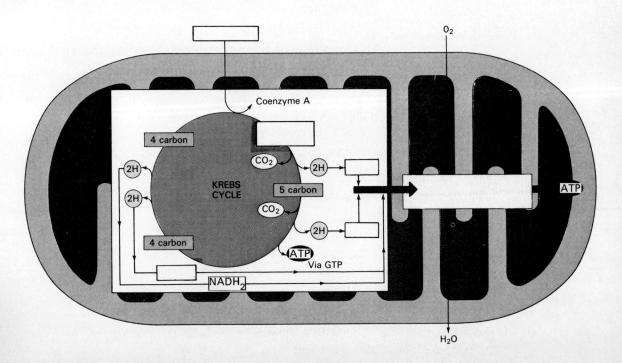

Figure 3.3

47. Using the terms listed below, complete the statements which follow.

    acetyl CoA   hydrogen   carbon dioxide   GTP   NADH2

    FADH2

    _____ is the two carbon compound that enters the Kreb's cycle

    _____ is the compound in which the carbon from acetyl CoA ends up after a revolution of the Kreb's cycle

    _____ which is removed from acetyl CoA is transferred to the electron transport system

    _____ will result in the production of 3 ATP molecules

    _____ will result in the production of 2 ATP molecules

    _____ is a high energy compound produced by substrate level phosphorylation during the Kreb's cycle

48. The catabolism of a simple carbohydrate molecule is known as _____.

49. _____ glycolysis is the complete oxidation of glucose to carbon dioxide and water.

50. _____ acid is the carbon-containing endproduct of glycolysis.

51. One molecule of glucose produces _____ molecules of pyruvic acid.

52. Inside of the mitochondria each pyruvic acid molecule loses a carbon in the form of carbon dioxide and is then converted into _____.

53. Complete the following table by writing in the number of ATP produced during each phase of aerobic glycolysis. Refer to Table 3.4 of your textbook for assistance.

AEROBIC GLYCOLYSIS

Amount of ATP   Source
(per glucose molecule)

_____ substrate level phosphorylation during glycolysis

_____ substrate level phosphorylation during the Kreb's cycle

_____ 2 FADH2 from the Kreb's cycle

_____ 2 NADH2 from glycolysis

_____ 2 NADH2 from the conversion of pyruvic acid into acetyl CoA

_____ 6 NADH2 from the Kreb's cycle

Subtotal _____

Losses   _____   conversion of glucose into glucose-6-phosphate and the initiation of glycolysis

Net Gain _____   from the complete oxidation of one glucose molecule to six carbon dioxide molecules and six water molecules

54. In the absence of oxygen, ATP can be generated for a limited time by means of _____ glycolysis.

55. During anaerobic glycolysis pyruvate is converted into _____.

56. Aerobic glycolysis is _____ times as efficient as anaerobic glycolysis.

57. During catabolism of triglycerides, glycerol is converted into _____ and routed into the Kreb's cycle.

58. Fatty acids are converted into acetyl CoA by the process of _____.

59. The passing of an amino group from one amino acid to another carbon chain is known as _____.

60. _____ is the removal of an amino group from an amino acid.

61. The carbon chain left by deamination enters the _____ and is catabol-ized.

62. The liver converts highly toxic ammonia which is derived from deamination into less toxic _____.

63. _____ is the process by which glucose is synthesized from three carbon intermediates.

64. Glucose molecules can be used to synthesize _____ which is stored in the liver and skeletal muscles.

65. Three fatty acids that cannot be synthesized and must be in the diet are _____, _____, and _____.

66. There are _____ essential amino acids that must be included in the diet.

67. The genetic code is contained in the _____ molecule.

68. Each genetic code word consists of three nucleotides in a row and codes for a (an) _____.

69. A _____ codes for a specific polypeptide chain.

70. The copying of the genetic code of DNA into mRNA is known as _____.

71. A triplet code word in mRNA is termed a _____.

72. _____ is the process of polypeptide construction from the information contained in mRNA.

73. The _____ of tRNA locks on to the appropriate codon of mRNA.

74. The _____ carried by each tRNA forms a peptide bond with the one that preceded it.

75. The triplet code words in the DNA molecules are CCC AAA TTT GGG. The codon of each in mRNA will be _____. The anticodon for each one in tRNA will be _____.

76. The enzyme which transcribes RNA is _____.

77. _____ are proteins closely associated with DNA in the nucleus.

78. The coils of DNA and histones form structures known as _____.

79. Genetic activity may be controlled by _____, _____, or _____.

80. Listed below are events that occur during mitosis. In the space provided, write in the phase of mitosis in which each event occurs.

    _____ disappearance of the nuclear membrane

    _____ alignment of the chromosomes on the equator

    _____ separation of the chromatids

    _____ replication of the DNA

    _____ appearance of the spindle

    _____ duplication of the centrioles

    _____ division of the cell

    _____ uncoiling of the chromosomes

    _____ appearance of the duplicated chromosomes

81. The difference between a liver cell and a nerve cell is not what genes are present, but what genes are available for _____.

SELF TEST

Circle the correct answer to each question.

1. Most ATP production occurs in the
   a. ribosome.   b. ER.   c. mitochondria.   d. lysosome.   e. centrosome.

2. These organelles contain digestive enzymes. They are the
   a. ribosomes.   b. Golgi apparatus.   c. centrosomes.   d. lysosomes.   e. ER.

3. The organelles which function in protein synthesis are the
   a. ribosomes.   b. Golgi apparatus.   c. lysosomes.   d. peroxisomes.   e. ER.

4. Pure water is _____ to a typical cell.
   a. isotonic   b. hypotonic   c. hypertonic   d. isosmotic   e. hyperbaric

5. The movement of molecules from an area of greater concentration to an area of lesser concentration best defines
   a. osmosis.   b. diffusion.   c. active transport.   d. filtration.   e. pinocytosis.

6. Active transport always requires
   a. a concentration gradient.   b. energy.   c. water.   d. electrolytes.   e. none of the above

7. Oxidative phosphorylation takes place in the
   a. nucleus.   b. mitochondria.   c. ribosomes.   d. ER.   e. peroxisomes.

8. Which of the following coenzymes is not a hydrogen acceptor?
   a. FAD   b. NAD   c. coenzyme A   d. coenzyme Q   e. FMN

9. The final electron (hydrogen) acceptor in the respiratory chain is
   a. cytochrome.   b. coenzyme Q.   c. coenzyme A.   d. FAD.   e. oxygen.

10. For every pair of hydrogen picked up by NAD during the Kreb's cycle, _____ ATP will be generated by the electron transport system.
    a. 1   b. 2   c. 3   d. 4   e. 5

11. During glycolysis glucose is broken down into
    a. acetyl CoA.   b. pyruvic acid.   c. glucose-6-phosphate.   d. carbon dioxide.
    e. water.

12. Fatty acids are converted into acetyl CoA by a process called
    a. alpha oxidation.   b. the tricarboxylic acid cycle.   c. beta oxidation.
    d. gluconeogenesis.   e. none of the above

13. The equivalent anticodon of the DNA triplet code word TTT will be
    a. AAA.   b. TTT.   c. CCC.   d. UUU.   e. GGG.

14. Which of the following play a role in protein synthesis?
    a. mRNA.   b. rRNA.   c. tRNA.   d. DNA.   e. More than one of the above is correct.

15. During mitosis the spindle appears in
    a. interphase.   b. prophase.   c. metaphase.   d. anaphase.   e. telophase.

CHAPTER 4
Tissues

OVERVIEW

Although single cells can execute all of the basic processes of life, these processes can be carried out more efficiently by a multicellular organization. Multicellularity permits specialization and division of labor. Just as a building can be constructed more efficiently by specialized groups of workers, so can the functions of life be carried out more efficiently by specialized groups of cells. These cells that organize and work together to execute a special function are termed tissues, and they represent the fourth level of organization in the living world.

This chapter introduces you to the fundamental tissue types that occur in our bodies. There are only four such fundamental tissue types, but they can be put together in a multitude of ways to construct all of the organs of our bodies. In addition to the tissue types themselves, you will be introduced to various gland and membrane types which tissues form. You will also examine the inflammatory response, a homeostatic response of the tissues to damage. Finally there is a major essay on cancer, one of the most severe pathologies that afflict tissues.

CHAPTER OUTLINE

A. Introduction
B. Epithelia
   1. Functions of epithelia
   2. Specializations of epithelial cells
      a. Specializations of epithelial membranes
   3. A classification of epithelia

        4. Glandular epithelia
   C. Connective tissues
        1. Connective tissues proper
             a. Wandering cells
             b. The fibers
             c. Loose connective tissue
             d. Dense connective tissue
             e. Elastic and reticular connective tissues
        2. Blood and lymph
        3. Cartilage
             a. Types of cartilage
             b. Growth and repair of cartilage
        4. Bone
   D. Membranes
   E. Muscle tissue
   F. Neural tissue
   G. Homeostasis at the tissue level
        1. The inflammatory response
   H. Essay: tumors and cancer

LEARNING ACTIVITIES

Complete the following items by supplying the appropriate word or phrase.

1. _____ is the study of tissues.
2. The four primary types of tissues are _____, _____, _____, and _____.
3. Epithelia tissues always have a _____ surface.
4. As a rule, epithelial tissues have a large number of packed cells and very little _____ material.
5. Epithelial gland cells that release their product into the blood are _____ cells.
6. The four functions of epithelia are

   _____

   _____

   _____

   _____

7. _____ is a connection between cells where the membranes are fused together.
8. A layer of _____ separates cell membranes that are attached to one another by desmosomes.
9. _____ junctions rely solely on a layer of proteoglycan to hold cell membranes together.
10. One of the most common proteoglycans used to glue cell membranes together is _____.
11. Epithelia are always attached to a special _____.
12. A cell with _____ has 20 times the surface area of a cell that lacks them.

13. The two major characteristics used to classify epithelia are numbers of _____ and shape of the _____.

14. A _____ epithelium has one cell layer.

15. A _____ has two or more cell layers.

16. _____ epithelia are composed of flattened cells.

17. The epithelium made up of box shaped cells is the _____ epithelium.

18. Tall, slender cells compose _____ epithelia.

19. The epithelium that lines the body cavities is known as a _____, while the one which lines the circulatory system is termed an _____.

20. Match the epithelial type listed below with the statements that follow.

    a. simple squamous   b. stratified squamous   c. simple cuboidal   d. transitional
    e. simple columnar   f. pseudostratified columnar

    _____ functions in absorption and friction reduction
    _____ composes the endothelium
    _____ found where mechanical stress occurs
    _____ specializes in secretion
    _____ lines the urinary bladder
    _____ lines the mouth
    _____ lines the digestive tract
    _____ lines the lower respiratory tract

21. Multicellular exocrine glands are classified as _____ or _____.

22. _____ glands have both endocrine and exocrine functions.

23. _____ glands have their product released by exocytosis.

24. In _____ glands, part of the cytoplasm is lost with the gland secretion.

25. _____ glands release their secretion by the bursting and subsequent death of the cells.

26. _____ are unicellular exocrine glands.

27. Label and color each of the epithelial tissues in Figure 4.1. Refer to Figures 4.4 - 4.7 in your textbook for assistance.

    simple squamous,   stratified squamous,   cuboidal, columnar,
    pseudostratified columnar,   transitional

28. The ducts of _____ exocrine glands branch repeatedly.

29. _____ glands secrete a watery solution containing enzymes.

30. All connective tissues contain _____, _____, and _____.

31. The two categories of cells found in connective tissue proper are _____ and _____.

32. The cells which synthesize connective tissue fibers are the _____.

33. Adipocytes are _____ cells.

34. Pigment-containing cells are termed _____.

35. Cells which phagocytize dead cells and pathogens are the _____.

36. The cytoplasm of mast cells contains the vasoactive chemicals _____ and _____.

37. The three basic fibers of connective tissues are _____, _____, and _____.

Label: simple squamous, stratified squamous, cuboidal, columnar, pseudostratified columnar, transitional

a—_____   b—_____

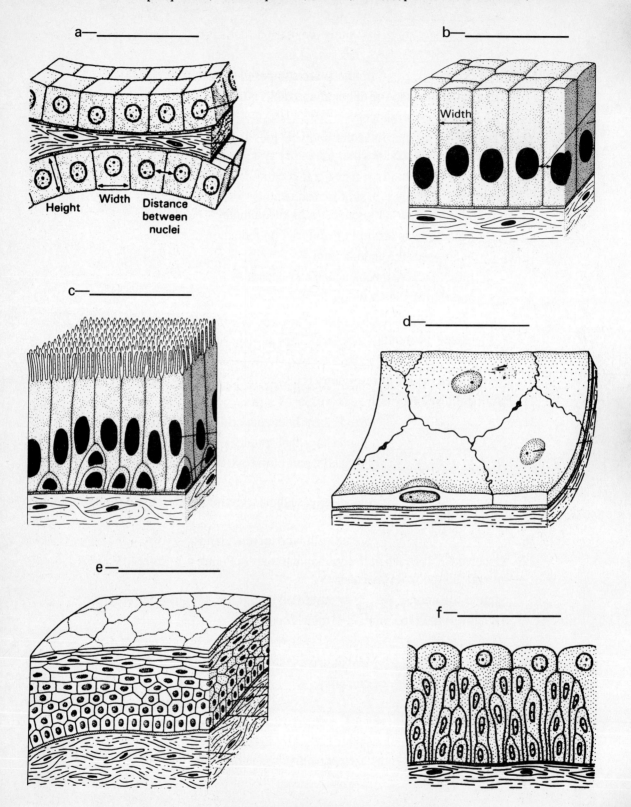

c—_____

d—_____

e—_____

f—_____

Figure 4.1

38. _____ connective tissue is the packing material of the body.

39.   Loose or areolar connective tissue can become _____ tissue.

40. Dense regular connective tissue makes up _____, _____, and _____.

41. In dense _____ connective tissue, the collagenous fibers show no definite orientation.

42. _____ connective tissue is dominated by elastic fibers.

43. The framework of the liver, spleen, and lymph nodes is formed by _____ connective tissue.

44. The connective tissues with a fluid matrix are _____ and _____.

45. The ground substance, or matrix of cartilage is composed of the proteoglycan called _____.

46. Because the matrix, of cartilage is solid, the cells lie in hollow pockets called _____.

47. The cells of cartilage are the _____.

48. As cartilage is avascular, all nourishment is provided by the _____, which is a surrounding membrane.

49. The three major types of cartilage are _____, _____, and _____

50. Label and color all of the structures in Figure 4.2, connective tissue proper, using the terms listed below. Refer to Figure 4.10 of your textbook for assistance.

reticular fibers, fixed macrophages, plasma cells, red blood cells, fat cell, blood vessel, mast cell, elastic fiber, free macrophage, collagen fiber, fibroblast, lymphocyte

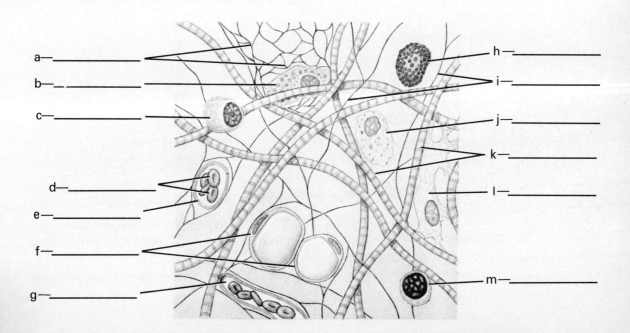

Figure 4.2

51. The cartilage found between the vertebrae is _____.
52. The surfaces of bones at joints is covered by _____ cartilage.
53. The outer ear contains _____ cartilage.
54. The two growth patterns found in cartilage are _____ and _____.
55. In bone _____ are organized around collagenous fibers.
56. The lacunae of bone contain the _____.
57. _____ are extensions of osteocytes that allow them to communicate with blood vessels and each other.
58. The three types of epithelial membranes are _____, _____, and _____.
59. _____ membrane lines the body cavities that open to the outside.
60. The closed, internal body cavities are lined by _____ membranes.
61. The _____ is a serous membrane that lines the abdominal cavity.
62. The _____ portion of a serous membrane lines the cavity while the _____ layer covers the organs.
63. The primary function of the serous membranes is to _____ friction.
64. The _____ membrane includes the stratified epidermis that covers the skin.
65. _____ membranes are composed entirely of connective tissues.
66. _____ membranes line the capsule that surrounds movable joints and produces a fluid that lubricates the joint.
67. The three types of muscle are _____, _____, and _____.
68. Skeletal muscle cells are multinucleate and have prominent _____.
69. _____ muscle cells form branching networks.
70. _____ muscle cells are small, spindle shaped, and possess a single nucleus.
71. The muscle type with the largest cells is _____ muscle.
72. _____ tissue is specialized for the conduction of impulses.
73. The basic nerve cell is the _____.
74. The supporting cells of nervous tissue are the _____ cells.
75. The major processes of the neuron are the _____ and _____.
76. Tissue damage always triggers an _____ response.
77. At the beginning of inflammation _____ and _____ released by mast cells cause blood vessels to dilate.
78. Increased blood flow to the inflamed area causes the skin to turn _____.
79. The histamine released by mast cells increases capillary permeability and more fluid moves into the injured area resulting in _____.
80. The pressure on nerve endings caused by swelling of the inflamed area produces _____.
81. The clotting protein fibrinogen is converted into _____, which forms a framework around the inflamed area, retarding the spreading of the response.

82. Macrophages and _____ from the blood enter the damaged area and begin to clean up.

83. The principal microphage in the inflammatory response is the _____.

84. Dead and dying cells along with tissue fluid and cellular debris form _____.

85. An _____ is a collection of pus in an enclosed tissue space.

86. Pus in an abscess which is not absorbed by surrounding tissues becomes enclosed by connective tissue forming a _____.

SELF TEST

Circle the correct answer to each question.

1. The basic tissue type that covers surfaces and lines hollow organs is _____ tissue.

   a. epithelial   b. connective   c. muscle   d. neural   e. blood

2. Gland cells are derived from _____ tissue.

   a. adipose   b. loose connective   c. epithelial   d. muscle   e. dense connective

3. The urinary bladder is lined by _____ epithelium.

   a. cuboidal   b. simple squamous   c. stratified squamous   d. transitional   e. columnar

4. The gland type in which cells are destroyed upon secretion is the _____ gland.

   a. apocrine   b. merocrine   c. holocrine   d. endocrine   e. sweat

5. The respiratory tract is lined by _____ epithelium.

   a. simple squamous   b. stratified squamous   c. simple columnar   d. pseudo-stratified columnar   e. transitional

6. Connective tissues contain

   a. fixed cells.   b. wandering cells.   c. fibers.   d. ground substance   e. More than one of the above is correct.

7. Which of the following is not considered to be connective tissue proper?

   a. loose connective tissue   b. adipose tissue   c. dense regular tissue   d. dense irregular tissue   e. cartilage

8. The fiber which is strong, flexible, and not stretchable is

   a. collagenous.   b. elastic.   c. reticula.r   d. keratin.   e. none of the above.

9. The cartilage cell is the

   a. fibroblast.   b. chondrocyte.   c. perichondrium.   d. osteocyte.   e. osteoblast.

10. The basic fiber forming cell is the

    a. fibroblast.   b. chondrocyte.   c. osteocyte.   d. adipose cell.   e. osteoblast.

11. The cartilage found between the vertebrae is

    a. elastic.   b. hyaline.   c. fibrous.   d. osteous.   e. collagenous.

12. Bone cells are nourished by cytoplasmic extensions termed

    a. osteocytes.   b. neuroglia.   c. canaliculi.   d. perichondrium.   e. periosteum.

13. The membranes that line body cavities that open to the outside are
    a. serous. b. synovial. c. mucous. d. fibrous. e. connective.
14. The support cells of nervous tissue are _____ cells.
    a. neurons b. striated c. neuroglial d. soma e. fibroblasts
15. The cell which releases the chemical substances that initiate inflammation is the
    a. macrophage. b. fibroblast. c. mast cell. d. plasma cell. e. microphage.

CHAPTER 5
The Integument:
Tissues in Combinations

OVERVIEW

The integumentary system consists of the skin and accessory structures such as hair and nails. It is the first organ and system to be surveyed because it clearly demonstrates how tissues interact to form organs, the next level of organization. In this chapter you will be introduced to the structure and function of the skin and its major derivatives.

As the skin is the principal organ which meets the environment, it is not surprising that it can suffer many indignities. Consequently, at the end of this chapter, a rather extensive examination is made of the clinical problems associated with the skin.

CHAPTER OUTLINE

      2. Structural integration
      3. Inflammation and repair
  D. Essay: Burns
  E. Clinical patterns
      1. Trauma
      2. Disorders of keratin production
      3. Invasion by microorganisms
      4. Abnormal pigmentation
      5. Integumentary inflammation
      6. Symptoms and diagnosis
      7. Treatment

LEARNING ACTIVITIES

Complete the following items by supplying the appropriate word or phrase.

1. _____ are combinations of tissues that perform complex functions.

2. The functions of the integument include:

    _____

    _____

    _____

    _____

    _____

    _____

3. The functional components of the integument are the _____ membrane and the _____ structures.

4. The two components of the cutaneous membrane are the _____ and the _____.

5. Match the epidermal layers listed below with the appropriate statement that follows.
   a. stratum germinativum   b. stratum spinosum   c. stratum granulosum   d. stratum lucidum   e. stratum corneum

   _____ The outermost layer of cells.

   _____ The layer which is the source of new epidermal cells.

   _____ The layer in which keratohyalin synthesis begins.

   _____ So named because the cells look like miniature pincushions.

   _____ Contains the cells that are filled with eleidin.

   _____ Dead layer of cells that are filled with keratin.

6. Keratohyalin is converted into _____ which in turn is transformed into keratin.

7. It takes approximately _____ days for a cell to move from the stratum germinativum to the stratum corneum.

8. For Figure 5.1, label all of the structures listed below. Color each of the listed structures. Refer to Figure 5.1 in your textbook for assistance.

hair shaft	nerve	sebaceous gland	erector muscle
sweat duct	hair follicle	sweat gland	blood vessel
fat	subcutaneous layer	dermis	epidermis

9. Thin skin is usually _____ mm in thickness.

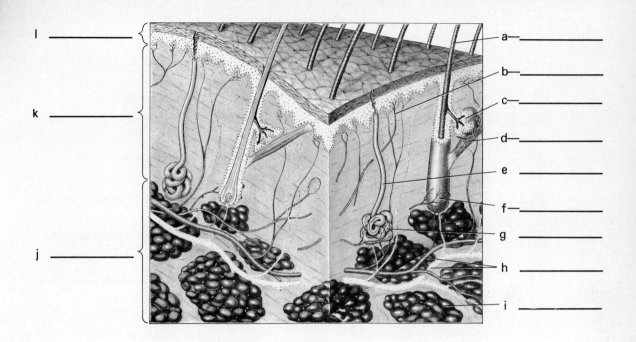

Figure 5.1

10. The _____ ridges extend into the dermis, increasing the area of contact between the two regions.

11. The epidermal ridge patterns are expressed at the surface as _____ prints.

12. Skin color represents an interplay of blood flow, carotene, and _____.

13. Melanocytes are found in the stratum _____.

14. Ultraviolet radiation causes an increase in _____ production.

15. Overexposure to ultraviolet light causes premature wrinkling of the skin and skin _____.

16. The dermis is made up of two layers, the _____ and the _____ layer.

17. Dermal _____ project between the epidermal ridges.

18. The reticular layer is composed of dense _____ connective tissue.

19. Skin wrinkles result from a reduction in the amount of _____ in the dermis.

20. A cancer of epithelial tissue is termed a _____.

21. Prolonged exposure to the sun can cause a _____ cell carcinoma.

22. Benign tumors of melanocytes produce _____.

23. Benign tumors of the dermis are termed _____.

24. Each body hair is contained in a hair _____.

25. It is the cells of the _____ which surrounds the papilla that is responsible for the formation of the hair proper.

26. The soft core of the hair shaft is the _____, while the stiff, outer layer is the _____.

27. The shaft is surrounded by a layer of hard keratin termed the _____.

28. The three types of hair on the body are _____, _____, and _____.

29. The _____ muscle can cause hair to stand up by pulling on the hair follicle.

30. Hair color is due to pigment production by the _____ in the papilla.

31. White hair results from _____ bubbles in the hair shaft.

32. _____ refers to the growth of hair on women in patterns associated with men.

33. _____ glands discharge a waxy secretion into the hair follicles.

34. The secretion of the sebaceous glands is known as _____.

35. _____ sweat glands are found in the armpit, around nipples, and in the groin.

36. The sweat glands which function in temperature regulation are the _____ sweat glands.

37. Perspiration which is mediated by sweat glands is known as _____ perspiration.

38. The _____ are modified sweat glands that produce earwax.

39. The body of the nail covers the _____, but growth occurs at the nail _____.

40. The stratum corneum that extends over the nail nearest its root is the _____ or _____.

41. The thickened stratum corneum that lies beneath the free edge of the nail is the _____.

42. _____ coordination of blood flow and sweat production plays an important part in thermoregulation.

43. Fibers of the reticular layer of the dermis are continuous with those of the _____.

44. The distribution of subcutaneous _____ tissue differs in men and women.

45. If the protective barriers of the skin are crossed, mast cells initiate a protective _____ response.

46. Widespread inflammation of the connective tissues is termed _____.

47. _____ ulcers are due to restriction of blood flow to areas of the skin.

48. Match the trauma term listed below with the appropriate description which follows.

    a. open wound   b. abrasion   c. incisions   d. laceration   e. puncture wound
    f. avulsion   g. contusion

    _____ A closed wound caused by dermal bleeding.

    _____ Wounds resulting from slender, pointed objects.

    _____ Chunks of tissue are torn away by impact.

    _____ A jagged, irregular tear.

    _____ An injury which produces a break in the epithelium.

    _____ A linear cut produced by a sharp instrument.

    _____ A wound that results from scraping against a solid instrument.

49. Calluses and corns are an example of _____.

50. An abscess which results from bacterial invasion of a sebaceous gland is known as a _____.

51. Freckles are an example of _____, and are associated with high _____ activity.

52. Mild cases of _____ in adults result in dandruff.

53. Generalized inflammation of the skin is known as _____.

SELF TEST

Circle the correct answer to each question.

1. Which of the following is not a function of the skin?

   a. sensitivity to the environment   b. storage of glycogen   c. synthesis of vitamin D
   d. maintenance of body temperature   e. excretion of wastes

2. The epidermal layer that produces new cells is the stratum

   a. germinativum.   b. spinosum.   c. granulosum.   d. lucidum.   e. corneum.

3. Eleidin

   a. is derived from keratin.   b. is derived from keratohyalin.   c. represents the final
   content of the stratum corneum.   d. is a breakdown product of keratin.   e. none of
   the above

4. The outer epidermal layer is the stratum

   a. germinativum.   b. spinosum.   c. granulosum.   d. lucidum.   e. corneum.

5. Finger prints are due to the

   a. papilla.   b. hair follicles.   c. epidermal ridges.   d. sebaceous glands.   e. sweat
   glands.

6. Dark pigment is produced by cells known as

   a. fibroblasts.   b. osteocytes.   c. chondrocytes.   d. melanocytes.   e. none of the
   above

7. The pink skin of albinos is due to

   a. melanin.   b. carotene.   c. blood flow.   d. pink epidermal cells.   e. more than
   one of the above is correct

8. Nourishment to the hair follicle is provided by the

   a. cortex.   b. medulla.   c. cuticle.   d. papilla.   e. sebaceous gland.

9. In the hair shaft, hard keratin is found in the

   a. cortex.   b. medulla.   c. papilla.   d. sebaceous gland.   e. vellus.

10. The structures which are responsible for "goose bumps" are the

    a. sebaceous glands.   b. papilla.   c. epidermal ridges.   d. arrector pili.   e. none of
    the above

11. The product of the sebaceous glands is

    a. cerumen.   b. sebum.   c. eleidin.   d. keratin.   e. keratohyalin.

12. The bulk of the body's sweat glands are

    a. apocrine.   b. eccrine.   c. myoepithelial.   d. holocrine.   e. endocrine.

13. Another name for the nail cuticle is the

    a. nail bed.   b. nail fold.   c. hyponychium.   d. eponychium.   e. nail groove.

14. Cellulitis is inflammation of the

    a. epidermis.   b. epithelium.   c. muscle.   d. nerve.   e. connective tissues.

15. A skin trauma which is characterized by being a jagged irregular tear is a (an)

    a. puncture wound.   b. incision.   c. abrasion.   d. avulsion.   e. laceration.

CHAPTER 6
An Orientation
to the Human Body

OVERVIEW

In the preceding five chapters you have ascended the scale of organization from atoms to organs. This chapter completes the trip up this scale by introducing you to the organ systems of the human body, and provides an overview of the entire body. Future chapters will examine each of the organ systems in detail. This chapter simply introduces each one and establishes their interactions with one another.

Once the organ system survey has been completed you will be introduced to whole body anatomy, surface anatomy, terms of directions, planes of section, and other methods used by anatomists to describe the location of the parts of the body. This will permit you to understand and use precise anatomical terminology in describing the location of body parts.

CHAPTER OUTLINE

- A. Introduction
- B. The functions of the organ systems
  1. Integumentary system
  2. Skeletal system
  3. Muscular system
  4. Nervous system
  5. Endocrine system
  6. Cardiovascular system
  7. Lymphatic system

8. Respiratory system
9. Digestive system
10. Urinary system
11. Reproductive system
C. The development of organ systems
D. A frame of reference for anatomical studies
1. Superficial anatomy
   a. Anatomical landmarks
   b. Anatomical regions
2. Anatomical directions
   a. Sectional anatomy
      (1) Sectional anatomy and technology
      (2) Body cavities
      (3) Connective tissue organization
E. Essay: The language of anatomy and physiology

LEARNING ACTIVITIES

Complete the following items by supplying the appropriate word or phrase.

1. The four characteristics shared by all organ systems are

_____

_____

_____

_____

2. Match the organ system listed below with the appropriate statements that follow.

a. integumentary b. skeletal c. muscular d. nervous e. endocrine f. cardiovascular g. lymphatic h. respiratory i. digestive j. urinary k. reproductive

_____ plays a major role in thermoregulation

_____ organized into an axial and appendicular division

_____ consists of a central and peripheral division

_____ supports the body and protects other organs

_____ consists of skeletal muscle tissue

_____ major function is the transport of materials to and from the cells

_____ stores important minerals

_____ system which is structurally and functionally linked to the cardiovascular system

_____ functions in the generation of blood cells

_____ coordinates cell activities

_____ aids in regulating pH by means of buffers

_____ functions in the coordination and regulation of internal operations

_____ participates in thermoregulation by transporting heat within the body

_____ cells of this system carry oxygen and also aid in defense

_____ causes body movement and maintains body position

_____ responsible for sensation and the control of all other body organs

_____ hormones from this system enter directly into the blood

_____ plays the major role in defending against microbial invasion

_____ regulates the body's response to starvation, stress, and dehydration

_____ filters the blood

_____ produces gametes

_____ includes the tonsils, thymus gland, and spleen

_____ transports oxygen into the blood

_____ consists of the kidneys, ureters, bladder, and urethra

_____ major component is a long tube

_____ returns interstitial fluid to the circulation

_____ pH balance in the body fluids is partially maintained by controlling carbon dioxide levels with this system

_____ accessory glands that secrete enzymes into this system

_____ excretes metabolic wastes

_____ absorbs nutrients and water into the blood

3. The digestive, respiratory, urinary, and reproductive systems all have passageways lined by _____.

4. The three primary germ or tissue layers of the developing embryo are the _____, _____, and _____.

5. The standard anatomical position has the body _____.

6. The mammary glands are located on the _____ of the trunk.

7. The vertebral column is _____ with respect to the digestive tract.

8. The chin is _____ to the nose.

9. The nose is _____ with respect to the ears.

10. The ribs are _____ with respect to the sternum.

11. The elbow is _____ with respect to the wrist.

12. The wrist is _____ with respect to the elbow.

13. The eyes are located on the _____ surface of the head.

14. Complete Figure 6.1 by supplying the names of the missing regions. Color each region. Refer to Figure 6.11 of your textbook for assistance.

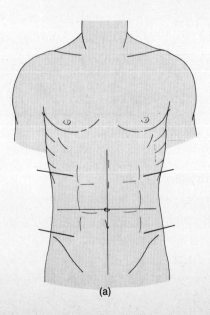
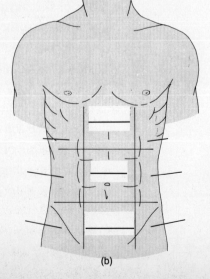

(a)   (b)

Figure 6.1

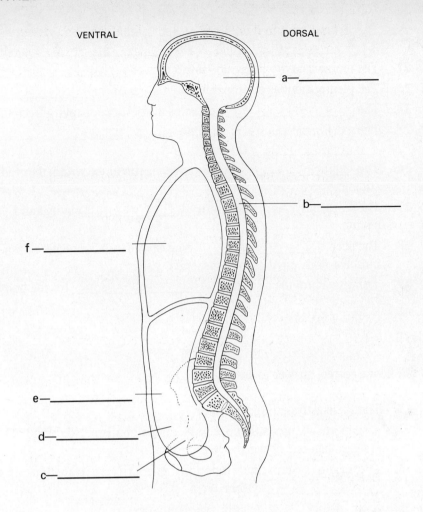

VENTRAL   DORSAL

a—_____

b—_____

f—_____

e—_____

d—_____

c—_____

Figure 6.2

15. Label Figure 6.2 using the terms listed below. Color each of the body cavities. Refer to Figure 5.14 of your textbook for assistance.

   cranial cavity   spinal cavity   thoracic cavity
   abdominopelvic cavity   abdominal cavity   pelvic cavity

16. A loaf of bread is sliced by making _____ sections through it.

17. A _____ section would divide the body into left and right halves.

18. A _____ section would divide the body into anterior and posterior portions.

19. Match the terms listed below with the appropriate statements that follow.

   a. PET   b. X-ray   c. Computerized tomography (CT)   d. DSR   e. MRI
   f. Ultrasound

   _____ used on pregnant women

   _____ utilizes radioisotopes in conjunction with a computer

   _____ blocked by structures of high radiodensity

   _____ using a moving X-ray tube combined with a computer, this technique permits views of body sections

   _____ utilizes a magnetic field to create images

   _____ permits the visualization of moving organs

20. The ventral body cavity is also known as the _____.

21. The ventral body cavity is subdivided into three chambers, the _____, the _____, and the _____ cavity.

22. The thoracic cavity is divided into a pair of _____ cavities.

23. The pleural cavities are separated by tissues of the _____.

24. The _____ separates the peritoneal cavity from the thoracic cavity.

25. The peritoneal cavity is divided into a _____ cavity and an _____ cavity.

26. The _____ body cavity is contained within the cranium and vertebral column.

27. The dorsal cavity is divided into a _____ cavity and a _____ cavity.

28. The deep _____ consist of dense connective tissue and it surrounds muscles, body cavities, and visceral organs.

29. The fascia provide a fibrous _____ for the body.

SELF TEST

Circle the correct answer to each question.

1. The skeletal system functions in
   a. support.   b. protection.   c. mineral metabolism.   d. blood cell generation.
   e. more than one of the above is correct

2. The organ system that functions primarily in transport of materials is the _____ system.
   a. circulatory   b. lymphatic   c. respiratory   d. digestive   e. urinary

3. The respiratory system functions in
   a. oxygenation of the blood.   b. pH balance.   c. carbon dioxide removal.   d. thermoregulation.   e. more than one of the above is correct

4. One of the major functions of the lymphatic system is
   a. pH balance.   b. return of interstitial fluid to the circulation.   c. generation of red blood cells.   d. excretion.   e. electrolyte balance.

5. The system that is responsible for excretion of most metabolic wastes is the
   a. lymphatic system.   b. digestive system.   c. urinary system.   d. respiratory system.   e. cardiovascular system.

6. The body surface region immediately inferior to the epigastric region is the
   a. hypochondriac region.   b. hypogastric region.   c. umbilical region.   d. lumbar region.   e. iliac region.

7. The eyes are superior and _____ with respect to the nose.
   a. medial   b. proximal   c. distal   d. lateral   e. dorsal

8. The finger tips are _____ with respect to the wrist.
   a. proximal   b. distal   c. lateral   d. inferior   e. superior

9. The feet are _____ with respect to the knees.
   a. inferior   b. superior   c. proximal   d. medial   e. lateral

10. An executioner's axe would make a _____ section through a victim's neck.
    a. sagittal   b. coronal   c. frontal   d. transverse   e. none of the above

11. A section that produced left and right halves of the body would be a _____ section.

   a. sagittal   b. coronal   c. frontal   d. transverse   e. inverse

12. The anatomical imaging system that utilizes radioactive isotopes is

   a. DSR.   b. CT.   c. PET.   d. MRI.   e. ultrasound.

13. The anatomical imaging system that utilizes a magnetic field is

   a. DSR.   b. CT.   c. PET.   d. MRI.   e. ultrasound.

14. Which of the following can be found in the ventral body cavity?

   a. pleural cavity   b. pericardial cavity   c. pelvic cavity   d. abdominal cavity   e. more than one of the above is correct

15. The sheets of dense connective tissue that provide a fibrous framework for the body are known as

   a. epithelial membranes.   b. synovial membranes.   c. joint capsules.   d. fascia.   e. none of the above

CHAPTER 7
Osseous Tissue

OVERVIEW

This chapter begins the second unit of your text, support and movement of the body. The functions of support and movement are largely functions of the skeletal and muscular systems. The first three chapters of this unit are concerned with the skeletal system. This chapter introduces you to the structure and function of bone as a tissue, whole bone anatomy, and outlines articulations between the bones. You will see how the two types of bone (spongy and compact) are organized, and how this difference in organization leads to differences in function. You will then be introduced to the growth and development of bone, and you will see how the skeletal system contributes to homeostasis. Having examined the histological aspects of the skeleton you will then begin a consideration of whole bone anatomy, and finally, there will be a survey of the major types of joints (articulations) that occur in the skeleton.

CHAPTER OUTLINE

    A. Introduction
    B. Internal organization
        1. Histological differences between compact and spongy bone
        2. Functional differences between compact and spongy bone
        3. The periosteum and endosteum
    C. Development and growth
        1. Intramembranous ossification

    2. Endochondral ossification
  D. Remodeling and homeostatic mechanisms
    1. The skeleton as a mineral reserve
      a. The importance of calcium ions
      b. The regulation of calcium ion concentration
    2. Injury and repair
    3. Clinical comment: a classification of fractures
    4. Clinical comment: stimulation of bone growth
  E. Anatomy of skeletal elements
    1. Clinical comment: heterotopic bones
    2. Bone markings
  F. Articulations
    1. Immovable joints (synarthroses)
    2. Slightly movable joints (amphiarthroses)
    3. Freely movable joints (diarthroses)
    4. Clinical comment: Rheumatism, arthritis, and synovial function
    5. Articular form and function
      a. Planes of motion
      b. A classification of synovial joints
        (1) Monaxial joints
        (2) Biaxial joints
        (3) Triaxial joints
    6. Describing dynamic motion
  G. Clinical patterns

LEARNING ACTIVITIES

Complete the following items by supplying the appropriate word or phrase.

1. The five functions of the skeletal system are:

    _____

    _____

    _____

    _____

    _____

2. Crystals of _____ account for almost two thirds of the weight of bone.

3. The collagen fibers of bone provide it with _____.

4. The two basic types of bone are _____ and _____.

5. The basic unit of compact bone is the _____.

6. The central canal of the osteon usually contains one or more _____.

7. The _____ connect the osteocytes to the blood vessels of the central canal.

8. The _____ are narrow sheets of calcified matrix that lie between the lacunae.

9. The spaces between the osteons of compact bone are filled by _____ lamellae.

10. Label Figure 7.1, the Haversian system of compact bone, using the terms listed below. Color the different structures. Refer to Figure 7.1 of your textbook for assistance.

    Haversian (central) canal   blood vessel   canaliculi
    lamella   osteocyte

11. Compact bone covers the _____ of bony elements except within the joint capsule.

12. _____ bone is found where bones are not heavily stressed.

13. In spongy bone the lamellae are organized into thin struts known as _____.

14. In a typical long bone, the diaphysis is composed of _____ bone and the epiphyses are made up of _____ bone.

15. The spaces between the trabeculae of spongy bone are filled with _____.

16. The periosteum of bone functions to isolate the bone, provide a route for nervous and circulatory supply, and participates in bone _____ and _____.

17. The periosteum becomes continuous with the collagen fibers of the joint _____.

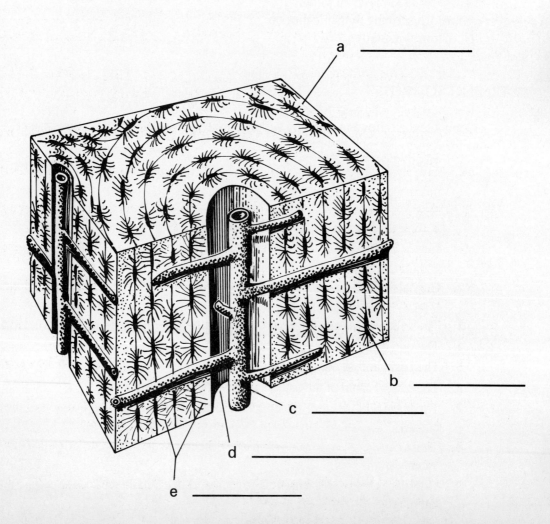

a _____

b _____

c _____

d _____

e _____

Figure 7.1

18. The _____ lines the marrow cavity, covers the trabeculae, and lines the central canals.

19. _____ are giant, multinucleated cells that are found in areas where the matrix of bone is exposed through the endosteum.

20. Osteoclasts function to _____ the bony matrix and thereby release stored minerals.

21. _____ is important in the regulation of calcium and phosphate concentrations in body fluids.

22. _____ are the cells that actively participate in osteogenesis.

23. In intramembranous ossification, the osteoblasts are located inside of a _____ tissue.

24. Because most intramembranous bone forms in the deep layers of the dermis, it is sometimes termed _____ bone.

25. Bones formed by intramembranous ossification include the roof of the _____, the _____, and the _____.

26. In the cartilage bone model, calcification leads to the death of the _____.

27. At the time that the chondrocytes of the bone model are dying, the perichondrium surrounding it converts into a _____.

28. The cells of the inner layer of the periosteum develop into _____ and produce a superficial layer of bone around the shaft.

29. As the decaying cartilage is broken down, it is replaced by _____.

30. Bone development proceeds from the center of _____ towards the ends of the cartilage model.

31. Initially the entire diaphysis is filled with _____ bone.

32. _____ erode the spongy bone in the center of the diaphysis and form the marrow cavity.

33. The blood vessels that lie in the center of osteons are connected to other blood vessels by _____ canals.

34. The collagenous fibers of tendons are cemented into the superficial _____ by osteoblasts.

35. Bone grows in diameter by the addition of bone to the outside by the osteoblasts and the balanced removal of bone from the inside by the _____.

36. The region between the diaphysis and epiphyses where cartilage continues to grow is the _____.

37. The continuous growth and replacement of cartilage in the metaphyses permits the bone to grow in _____.

38. The thin layer of cartilage that continues to grow in the metaphyses is termed the _____ plate.

39. _____ and _____ are two hormones that maintain the normal activity of the epiphyseal plates.

40. Once the epiphyseal plates ossify, growth in length _____.

41. An individual can grow no taller after the epiphyseal plates have _____.

42. Up to _____ percent of the mineral content of bone is removed and replaced each year.

43. Stress on bone results in the generation of minute _____ fields by the crystals which in turn attract osteoblasts.

44. The accumulation of osteoblasts at the site of bone stress results in _____ of new bone.

45. Heavily stressed bones become _____ and _____.

46. Regular _____ is important for the maintenance of normal bone structure.

47. Calcium ions are especially important to the normal functioning of _____ and _____ cells.

48. When the calcium level in the blood rises above normal the thyroid gland produces the hormone _____.

49. The overall effect of calcitonin is to _____ serum calcium levels.

50. If the calcium concentrations fall below normal, the parathyroid glands produce _____ hormone.

51. The overall effect of parathyroid hormone is to _____ serum calcium levels.

52. Complete Figure 7.2 by supplying the missing information. Refer to Figure 7.8 of your textbook for assistance.

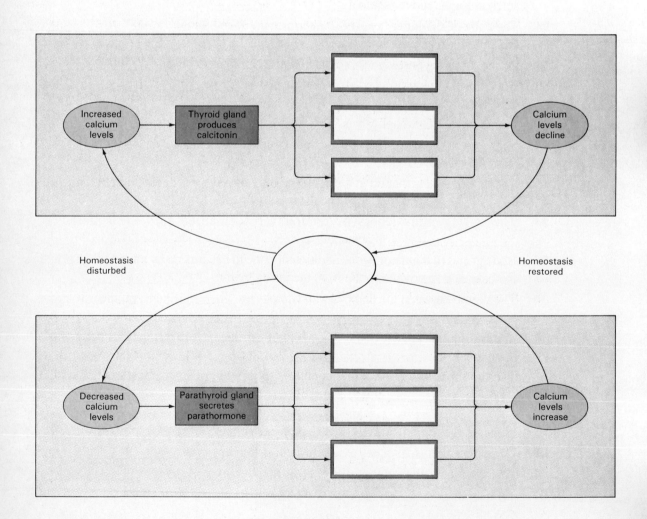

Figure 7.2

53. Match the fracture type listed below with the appropriate descriptive statement which follows.

a. simple b. compound c. comminuted d. nondisplaced e. impacted
f. transverse g. greenstick h. spiral i. Colles' j. Pott's k. compression
l. epiphyseal

_____ occurs in vertebrae which are subjected to severe stress

_____ occurs along the line where the matrix is calcifying

_____ type of fracture that is completely internal

_____ bony elements retain their alignment

_____ break in the distal portion of the radius

_____ the bone breaks across the long axis

_____ fracture usually seen in children whose long bones have not yet ossified completely

_____ occurs in the ankle and affects both lower leg bones

_____ one bone fragment is driven into another

_____ fracture produced by twisting stresses

_____ broken end of a bone protrudes through the skin

54. Following a bone break, a fibrous meshwork forms in the damaged area as a result of the fracture _____.

55. Following a break, the cells of the periosteum and endosteum rapidly proliferate and form an external _____ around the damaged area.

56. An internal _____ organizes within the marrow cavity and between the broken ends of the shaft.

57. The calluses are formed initially in _____, which is later replaced by bone.

58. Following ossification of the calluses, _____ and _____ will remodel the region.

59. The six major classes of bones based upon shape and/or origin are

_____, _____, _____,

_____, _____, and _____.

60. The _____ are sesamoid bones possessed by everyone.

61. Bones that develop in unusual places are termed _____.

62. In the condition _____ bone begins to form around the skeletal muscles.

63. Complete the following chart by writing the correct term. Refer to Table 7.2 of your text for assistance.

_____ projection or bump

_____ large, rough projection

_____ smaller, rough projection

_____ small, rounded projection

_____ prominent ridge

_____ low ridge

_____ expanded articular end of an epiphysis

_____ smooth, rounded, or oval articular process

_____ smooth, grooved articular process which is shaped like a pulley

_____ small, flat articular surface
_____ pointed process
_____ shallow depression
_____ narrow groove
_____ deep pocket or socket
_____ rounded passageway for blood vessels and/or nerves
_____ an elongate cleft
_____ canal leading through the substance of bone
_____ chamber within a bone

64. An _____ occurs wherever two bones interact.

65. The three major types of articulations found in the body are _____, _____, and _____.

66. _____ are synarthrotic joints between the bones of the skull.

67. Suture joints are bound together by dense _____ tissue.

68. The joint between each tooth and its socket is termed a _____.

69. The _____ plate is an example of an immovable joint connected by cartilage.

70. A _____ joint is an amphiarthrosis held together by collagenous fibers.

71. A _____ joint is an amphiarthrosis held together by a pad of fibrocartilage.

72. Synovial joints are typically found at the ends of _____ bones.

73. The end of each bone at a synovial joint is covered by a thin layer of articular _____.

74. The synovial joint is surrounded by a fibrous joint _____.

75. The articular cartilage functions to _____ friction.

76. The joint capsule is reinforced by _____.

77. Complex joints such as the knee may have additional pads of cartilage that subdivide the synovial cavity. These pads are termed _____.

78. Small pockets of synovial fluid that reduce friction between structures that rub against each other are termed _____.

79. A _____ occurs when the articular surfaces of the bones in a synovial joint are displaced.

80. _____ is the fusion of bones that occurs in a joint following extensive increases in friction.

81. _____ is a general term for pain and stiffness of joints.

82. _____ is usually found in older persons and results from the cumulative wear and tear on the joint surfaces.

83. _____ arthritis is an inflammatory condition.

84. The three possible types of movement are _____, _____, and _____.

85. The two types of monaxial joints are _____ and _____.

86. _____ joints permit rotation in one plane only.

87. _____ joints permit angular movement in one plane only.

88. The three types of biaxial joints are the _____, _____, and _____.

89. The fingers and toes are connected to the bones of the palms and soles by _____ joints.

90. _____ joints are known as ball and socket joints and permit movement in many different planes.

91. Match the type of movement listed below with the appropriate descriptive statement that follows.

a. flexion b. extension c. circumduction d. abduction e. adduction f. inversion g. eversion h. opposition i. pronation j. supination k. protraction l. retraction m. depression n. elevation

_____ occurs only with the thumb

_____ lowering of the jaw

_____ increases the angle between two bones

_____ decreases the angle between two bones

_____ movement of an arm or leg so that a cone is circumscribed

_____ movement of the ankle inward

_____ movement of the ankle outward

_____ movement of an appendage away from the midline of the body

_____ movement of an appendage towards the midline of the body

_____ movement of a structure forward in a horizontal plane

_____ movement of a structure backward in a horizontal plane

_____ rotation of the palms upward

_____ rotation of the palms downward

_____ raising of the jaw

92. Match the clinical condition listed below with the appropriate descriptive statement which follows.

a. hyperostosis b. osteopenia c. osteoporosis d. scurvy e. osteomalacia f. rickets g. osteogenesis imperfecta h. Marfan's syndrome i. achondroplasia j. gigantism k. osteosarcoma l. chondrosarcoma m. osteomyelitis

_____ results from a reduction in bone mass

_____ produced by a vitamin C deficiency

_____ results in softening of the bones

_____ form of osteomalacia seen in children due to vitamin D deficiency

_____ usually results from a bacterial invasion of bone

_____ results from an overproduction of growth hormone

_____ cancer of the periosteum or endosteum

_____ cancer of the epiphyseal or articular cartilages

_____ inadequate ossification

_____ excessive formation of bone

_____ inherited condition that affects the organization of collagen fibers, resulting in very fragile bones

_____ inherited condition of inactivity on the part of the epiphyseal plates

_____ inherited disorder that results in very long and slender arms and legs

SELF TEST

Circle the correct answer to each question.

1. Which of the following is a function of the skeletal system?

   a. support b. protection c. blood cell production d. mineral storage e. more than one of the above is correct

2. The structural unit of compact bone is the

   a. osteocyte. b. canaliculi. c. osteon. d. trabeculae. e. osteoclast.

3. The membrane that lines the marrow cavity is the

   a. perichondrium. b. endosteum. c. periosteum. d. synovial. e. none of the above

4. The cell that is responsible for osteogenesis is the

   a. osteoblast. b. osteocyte. c. osteoclast. d. mast cell. e. chondroblast.

5. Bone that forms in connective tissue membranes is termed

   a. endochondral. b. cartilage replacement. c. dermal. d. epidermal. e. none of the above

6. Which of the following bones would not be formed by endochondral osteogenesis?

   a. long bones of the leg b. long bones of the arms c. finger bones d. vertebrae e. cranial bones

7. Growth in length of the long bones is permitted because of the

   a. osteoclasts. b. osteocytes. c. trabeculae. d. epiphyseal plates. e. central canals.

8. The hormone which activates osteoclasts is

   a. calcitonin. b. parathyroid hormone. c. thyroxine. d. growth hormone. e. the sex hormones.

9. Following a fracture, a cartilaginous _____ forms between the broken ends.

   a. callus b. hematoma c. spongy bone d. membrane e. none of the above

10. Bones which form between the sutures of the skull bones are the _____ bones.

    a. irregular b. sesamoid c. wormian d. short e. flat

11. A blood vessel would pass through a bone via a (an)

    a. foramen. b. alveolus. c. condyle. d. facet. e. spine.

12. A joint in which the bones are connected by cartilage would be a

    a. suture. b. syndesmosis. c. synovial. d. synchondrosis. e. triaxial joint.

13. Which of the following are part of a typical synovial joint?

    a. capsule b. synovial membrane c. articular cartilage d. ligaments e. more than one of the above is correct

14. Movement of the arm away from the body to the side would be an example of

    a. flexion. b. extension. c. circumduction. d. abduction. e. adduction.

15. A fracture which resulted from a twisting stress would be a _____ type.

    a. comminuted b. transverse c. spiral d. ectopic e. nonunion

CHAPTER 8
The Skeletal System: Axial Division

OVERVIEW

The skeletal system is divided into two major divisions. The axial division consists of the long axis of the system, the skull, vertebral column, and thorax. The second division is the appendicular division which consists of the appendages and the girdles which tie the appendages to the axial division. This chapter is devoted to a consideration of the axial division. You will learn the major bones and their principal markings in this chapter. The skull is the most complex portion of the entire skeletal system and you will have to devote more time to it than to the other parts of the skeleton. When you examine the structure of the skull and vertebral column, keep in mind the protective function of the skeleton. Nowhere is this function more aptly demonstrated than here, where the entire central nervous system is enclosed by these remarkable structures. Likewise it requires very little imagination to see how the cage formed by the ribs, vertebral column, and sternum serve to protect the heart and lungs.

As a rule the skeletal system's anatomy is studied more closely and in more detail than any other system. The reason for this is that many parts of the other systems are keyed to the skeletal system. For example we have ulnar nerves, arteries, and veins. These are so named because they follow the ulnar bone of the forearm. Therefore, if you know your skeletal anatomy well, you can often determine where another structure is to be found simply based upon its name. Keep this in mind while learning the details of the skeletal system. A little extra effort here will pay big dividends later on in your study of anatomy.

CHAPTER OUTLINE

   A. Introduction
   B. The skull

1. The adult skull
   a. Superficial anatomy of the skull
   b. Sectional anatomy of the skull
   c. Individual bones of the skull
      (1) The occipital bone
      (2) The parietal bones
      (3) The frontal bone
      (4) The sphenoid bone
      (5) The maxillary bones
      (6) The nasal complex
      (7) The temporal bones
      (8) The mandible
      (9) The hyoid
2. The skull of infants and children

C. The neck and trunk
   1. Spinal curvature
   2. Introductory anatomy of the vertebral column
   3. Regional structure and function
      a. Cervical vertebrae
   4. The thorax
      a. Thoracic vertebrae
      b. The ribs and sternum
      c. Lumbar vertebrae
         (1) Problems with lumbar discs
      d. The sacrum and coccyx
D. Essay: A matter of perspective

LEARNING ACTIVITIES

Complete the following items by supplying the appropriate word or phrase.

1. The _____ division of the skeleton forms the longitudinal axis of the body.
2. There are _____ bones found in the axial division.
3. The _____ division consists of the appendages and girdles.
4. The appendicular division consists of _____ bones.
5. There are _____ bones in the skull, _____ in the cranium, and _____ associated with the face.
6. Match the suture listed below with the appropriate statement which follows.

   a. coronal b. sagittal c. lambdoidal d. occipitomastoid e. sphenosquamosal f. squamosal

   _____ articulation between the parietals and the occipital
   _____ where the frontal bone joins the parietals
   _____ separates the occipital from the temporal bone
   _____ articulation between the temporal bone and the parietal
   _____ where the sphenoid joins the temporal bone
   _____ articulation between the two parietal bones

7. In the spaces provided below, write the name of the bone which best fits the statement that follows.

_____ bone that surrounds the foramen magnum

_____ two bones that form the top part of the cranium

_____ bone that forms the anterior part of the calvarium

_____ bones form the bridge of the nose

_____ the upper jaw bones

_____ small, delicate bone that contributes to the medial portion of each orbit

_____ the upper jaw bones

_____ lower jaw bone

_____ cheek bone

_____ bone that contains the external auditory meatus

_____ bone that forms the inferior part of the nasal septum

_____ contains the perpendicular plate that forms the superior part of the nasal septum

_____ contains the sella turcica

_____ bones that form the posterior margin of the bony palate

8. There are six bones that contribute to the orbits. They are the _____, _____, _____, _____, _____, and _____.

9. The four bones that contain paranasal sinuses are the _____, _____, _____, and _____.

10. The anterior two thirds of the hard palate is formed by the _____ bones.

11. The sella turcica is a depression in the sphenoid bone in which the _____ gland rests.

12. The teeth are contained in the _____ processes.

13. The cribiform plate of the ethmoid permits the passage of the _____ nerves.

14. The superior and medial nasal conchae are projections into the nasal cavity from the ethmoid bone. The _____ nasal concha is a separate bone.

15. The _____ is suspended by ligaments from the styloid processes of the temporal bones.

16. Match the foramen in column B with the bones on which they reside in column A. Refer to Table 8.1 of your text for assistance.

A	B.
_____ mandible	a. foramen magnum
_____ temporal	b. hypoglossal canal
_____ maxilla	c. jugular foramen
_____ sphenoid	d. supraorbital foramen
_____ frontal	e. optic foramen
_____ occipital	f. superior orbital fissure
	g. foramen rotundum
	h. foramen ovale
	i. foramen lacerum
	j. intraorbital foramen
	k. carotid foramen

l. internal acoustic meatus

m. mental foramen

n. mandibular foramen

17. The fibrous areas between the cranial bones of infants are known as

_____.

18. The most significant growth of the skull occurs before the age of

_____.

19. Premature closure of one or more of the fontanelles results in the condition known as _____.

20. Failure of the two maxillary bones to fuse together properly results in a _____ palate.

21. Label Figure 8.1, a lateral view of the skull, using the terms listed below. Color each of the bones. Refer to Figure 8.2 of your textbook.

parietal	frontal	occipital	temporal	sphenoid
ethmoid	mandible	maxilla	zygomatic	lacrimal
nasal				

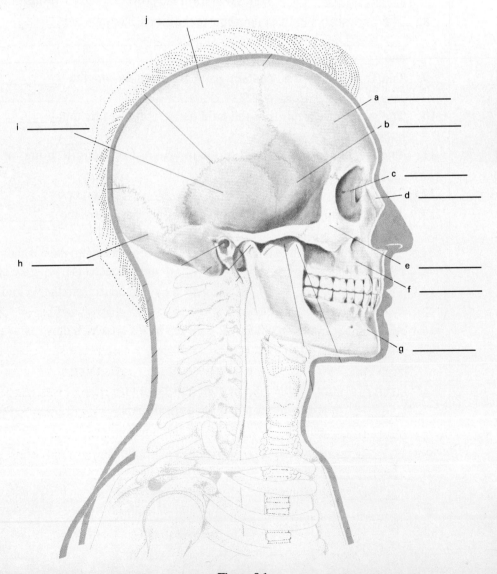

Figure 8.1

22. Label Figure 8.2, anterior and posterior views of the skull, using the terms listed below. Color each of the bones. Refer to Figure 8.3 of your textbook for assistance.

parietal	supraorbital foramen	sphenoid	temporal
ethmoid	lacrimal	zygomatic	nasal
inferior nasal concha	vomer	mental foramen	mandible
middle nasal concha	infraorbital foramen	coronal suture	maxilla
sagittal suture	occipital	lambdoidal	suture
mastoid process	occipitomastoid suture		

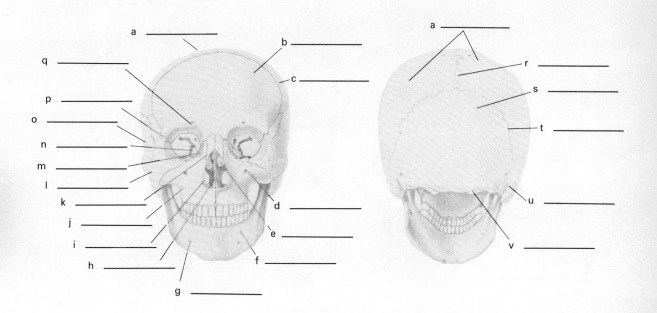

Figure 8.2

23. Label Figure 8.3, floor of the cranium, using the terms listed below. Color each of the bones. Refer to Figure 8.4 of your textbook for assistance.

crista galli	cribiform plate	anterior fossa	sphenoid
temporal	middle fossa	sella turcica	
posterior fossa	foramen magnum	internal acoustic canal	
jugular foramen	foramen lacerum	petrous portion of temporal bone	
foramen ovale	foramen rotundum	optic foramen	

24. Write the number of bones found in each region of the vertebral column in the space provided.

_____ cervical vertebrae

_____ thoracic vertebrae

_____ lumbar vertebrae

_____ sacrum

_____ coccyx

25. There are _____ normal spinal curves.

Label: crista galli cribiform plate anterior fossa sphenoid temporal middle fossa sella turcica
posterior fossa foramen magnum internal acoustic canal jugular foramen
foramen lacerum petrous portion of temporal bone foramen ovale foramen
rotundum optic foramen

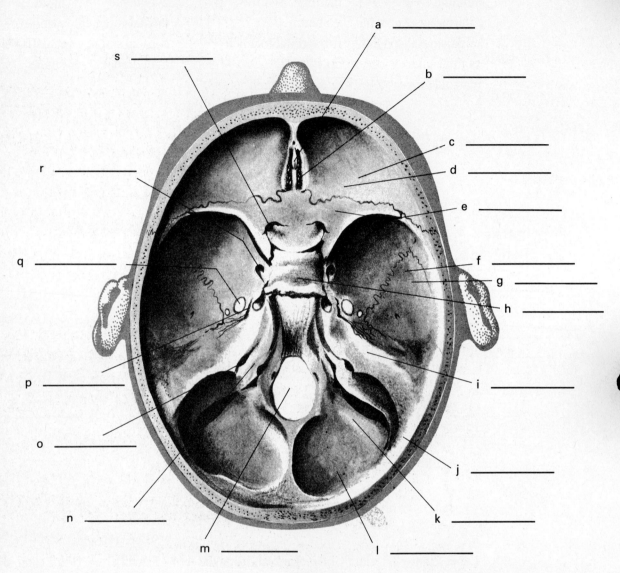

a _____
b _____
c _____
d _____
e _____
f _____
g _____
h _____
i _____
j _____
k _____
l _____
m _____
n _____
o _____
p _____
q _____
r _____
s _____

Figure 8.3

26. The mass of each vertebra is concentrated in its _____.

27. The outer walls of the vertebral foramen are formed by the _____.

28. The _____ extend from the pedicles and unite to form the spinous process.

29. The _____ processes project laterally from the pedicles.

30. Vertebrae articulate with each other by means of the superior and inferior _____ processes.

31. Adjacent centra are separated by an intervertebral _____.

32. The intervertebral discs function as _____ absorbers.

33. The first cervical vertebra is the _____, and the second cervical vertebra is the _____.

34. When we turn our heads sideways, we are utilizing the joint between the atlas and the _____.

35. Each thoracic vertebra articulates with a _____.

36. There are _____ pairs of ribs.

37. The first seven pairs of ribs are termed _____ because each connects to the sternum via its own separate cartilage.

38. Ribs 8 - 10 are termed _____ because they do not have their own individual cartilage connections to the sternum.

39. Ribs 11 and 12 are termed _____ because they have no connection with the sternum.

40. The three components of the sternum are the _____, _____, and _____.

41. The bulk of the weight of the upper body falls directly on the _____ region of the spinal column.

42. A _____ disc results in pain due to pressure on the spinal nerve roots.

43. The sacrum is formed from the fusion of _____ sacral vertebrae.

44. At the sacroiliac joint, the sacrum articulates with the _____ bones.

45. _____ results when the vertebral lamina fail to unite during development.

46. Label Figure 8.4, the vertebral column, using the terms listed below. Color each of the regions of the vertebral column with a different color. Refer to Figure 8.16 of your text for assistance.

cervical   thoracic   lumbar   sacral

47. Label Figure 8.5, representative vertebrae, using the terms listed below. Color each vertebra. Refer to Figures 8.18, 8.19, and 8.20 of your text for assistance.

transverse process   spinous process   pedicle   lamina

superior articular process   transverse foramen   centrum

vertebral foramen   inferior articular process rib   facet

48. An exaggerated curvature of the thoracic region is termed _____.

49. A lateral curvature anywhere in the vertebral column is known as _____.

50. An exaggerated curvature in the lumbar region results in _____.

SELF TEST

Circle the correct answer to each question.

1. The front part of the cranium is composed of the _____ bone(s).
    a. parietal   b. occipital   c. frontal   d. sphenoid   e. temporal

2. The cheek bone is the
    a. parietal.   b. temporal.   c. zygomatic.   d. sphenoid.   e. maxilla.

3. The superior and medial nasal conchae are parts of the _____ bone.
    a. sphenoid   b. ethmoid   c. zygomatic   d. maxilla   e. vomer

4. Which of the following bones does not contribute to the orbit?
    a. maxilla   b. parietal   c. sphenoid   d. frontal   e. lacrimal

5. The inferior part of the nasal septum is formed by the
    a. ethmoid.   b. vomer.   c. lacrimal.   d. frontal.   e. palatines.

6. The mental foramen is found on the
    a. maxilla.   b. mandible.   c. occipital.   d. sphenoid.   e. temporal.

Label: cervical   thoracic   lumbar   sacral

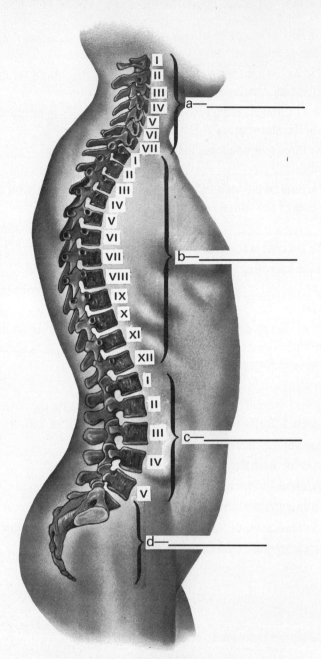

Figure 8.4

7. Which of the following is not found on the temporal bone?

   a. zygomatic process   b. mastoid process   c. external auditory meatus   d. petrous portion   e. temporal process

8. The frontal bone is separated from the parietals by the

   a. sagittal suture.   b. coronal suture.   c. squamosal suture.   d. lambdoidal suture. e. sphenosquamosal suture.

9. The fibrous areas between the cranial bones of infants are known as the

   a. fontanelles.   b. membrane areas.   c. soft spots.   d. congenital defects.   e. none of the above

10. There are _____ thoracic vertebrae.

    a. 5   b. 12   c. 6   d. 2   e. 1

Label: transverse process   spinous process   pedicle   lamina   superior articular process   transverse
foramen   centrum   vertebral foramen   inferior articular process rib   facet

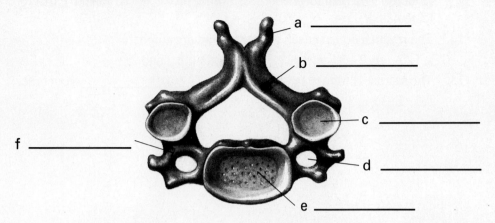

a _____

b _____

c _____

f _____

d _____

e _____

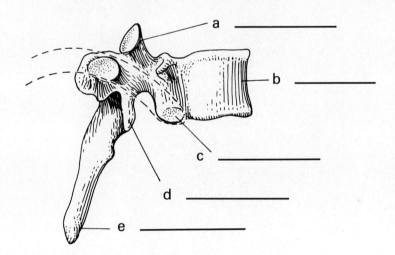

a _____

b _____

c _____

d _____

e _____

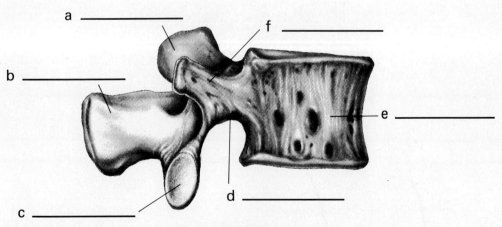

a _____

f _____

b _____

e _____

d _____

c _____

Figure 8.5

11. The spinous process is formed by the fusion of the
    a. lamina. b. pedicles. c. transverse processes. d. centra. e. intervertebral
    foramina.

12. A lateral curvature in the spinal column is termed
    a. lordosis. b. kyphosis. c. scoliosis. d. swayback. e. hunchback.

13. The most superior part of the sternum is the

a. body. b. manubrium. c. xiphoid process. d. costal cartilage. e. vertebrochondral rib.

14. The ribs that do not attach to the sternum are numbers

a. 1 - 7. b. 2 - 5. c. 8 - 12. d. 11 and 12. e. 1 and 2.

15. The sacrum is formed by the fusion of _____ vertebrae.

a. 1 b. 2 c. 3 d. 4 e. 5

CHAPTER 9
The Skeletal System: Appendicular Division

OVERVIEW

This chapter completes the discussion of the skeletal system. The appendicular division includes the bones of the appendages and the girdles. The appendages include the arms and the legs. The structural pattern of arms and legs is similar. There is a single large bone in the upper portion of each appendage and two smaller bones in the lower portion of each limb. The distal portion of each consists of a series of small bones with numerous joints, forming a foot in the case of the legs, and hands in the case of the arms. The girdles function to attach the appendages to the axial skeleton. The shoulder girdle (pectoral) consists of two bones, the clavicle and scapula. The hip girdle (pelvic) consists of a single bone (on each side) known as the coxa. Besides a consideration of the bones that make up the appendicular system, this chapter also includes a discussion of the principal synovial joints that occur in this division.

CHAPTER OUTLINE

A. Introduction
B. The pectoral girdle and arm
   1. The pectoral girdle
      a. The scapula
      b. The clavicle
   2. Functional anatomy of the shoulder
   3. The arm

LEARNING ACTIVITIES

Complete the following items by supplying the appropriate word or phrase.

1. The pectoral girdle consists of the _____ and the _____.

2. The point on the scapula that articulates with the humerus is the _____ fossa.

3. The _____ process of the scapula articulates with the clavicle.

4. In the clavicle, the _____ end is the larger.

5. A shoulder _____ is when the acromioclavicular joint undergoes dislocation.

6. The _____ joint permits the greatest range of motion of any joint in the body.

7. The cartilage that covers the glenoid fossa is known as the glenoid _____.

8. The articular capsule of the shoulder joint has a relatively _____ fit which permits an extensive range of motion.

9. The capsule of the shoulder joint is reinforced by the _____ ligaments.

10. The _____ ligament originates at the base of the coracoid process and inserts on the humerus.

11. The _____ of the shoulder play a larger role in stabilizing the shoulder joint than do the ligaments.

12. The upper arm bone is known as the _____.

13. The greater and lesser tubercles located on the head of the humerus serve as attachment points for _____.

14. The _____ neck of the humerus marks the distal limits of the articular capsule.

15. The _____ neck of the humerus corresponds to the metaphysis of the growing bone and is a common fracture site.

16. The deltoid muscle of the shoulder attaches to the _____ tuberosity located on the shaft of the humerus.

17. The _____ and _____ are articular surfaces located at the distal end of the humerus.

18. Label Figure 9.1, the pectoral girdle, using the terms listed below. Color each bone. Refer to Figures 9.1 and 9.2 of your textbook for assistance.

acromion   lateral border   medial border

subscapular fossa   superior angle   superior border

scapular notch   coracoid process   supraspinous fossa

infraspinous fossa inferior angle   humerus   glenoid fossa

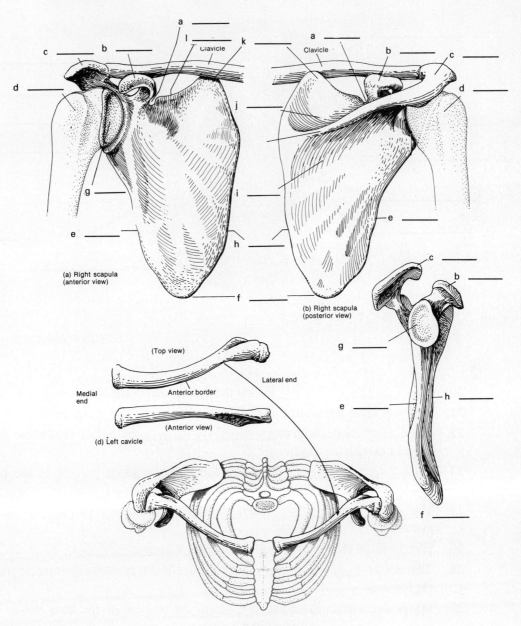

Figure 9.1

19. Label Figure 9.2, the shoulder joint, using the terms listed below. Color each labeled structure. Refer to Figure 9.4 of your text for assistance.

Clavicle   Coracoacromial ligament   Tendon of biceps

Subcoracoid bursa   Glenohumeral ligaments   Subscapular bursa

Subscapularis muscle   Glenoid labrum   Glenoid cavity

Articular capsule   Subacromial bursa   Acromion

Acromioclavicular ligament

Label: Clavicle   Coracoacromial ligament   Tendon of biceps   Subcoracoid bursa
Glenohumeral ligaments   Subscapular bursa   Subscapularis muscle
Glenoid labrum   Glenoid cavity   Articular capsule   Subacromial
bursa   Acromion   Acromioclavicular ligament

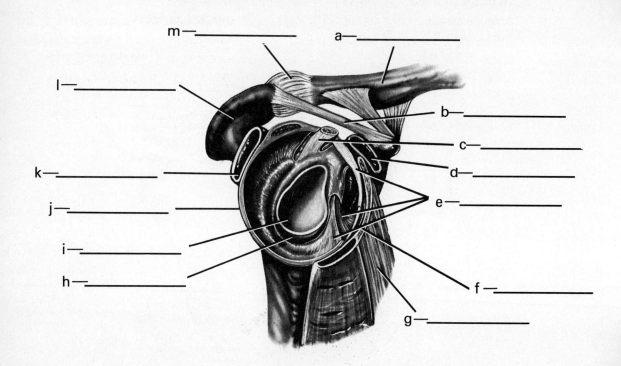

Figure 9.2

20. The _____ lies on the thumb side of the arm.

21. The radioulnar joints are a pair of _____ articulations.

22. The ulna articulates with the _____ of the humerus while the radius articulates with the capitulum.

23. The _____ process forms the superior portion of the proximal epiphysis of the ulna.

24. The _____ notch of the ulna articulates with the trochlea of the humerus.

25. The coronoid process is found on the _____.

26. The radial _____ is the attachment point for the muscle that flexes the forearm.

27. When the arm is extended, the olecranon process of the ulna drops into the _____ fossa of the humerus.

28. Pulling on the tendon that attaches to the ulnar tuberosity produces _____.

29. The elbow joint is extremely stable because the bony surfaces _____.

30. There are _____ carpal bones in the wrist.

31. The radiocarpal articulation is _____, but the intercarpal articulations are _____ joints.

32. The palm of the hand is composed of five _____ bones.

33. The carpal-metacarpal articulation at the thumb is a _____ joint.

34. The other carpal-metacarpal articulations are all _____ joints.

35. The thumb contains _____ phalanges while each remaining finger has three.

36. The phalangeal-metacarpal articulations are all _____ joints.

37. There are a total of _____ bones in the hand.

38. A major advantage of all the bones and joints found in the hand is the increased _____ that they permit.

39. Label Figure 9.3, the humerus, using the terms listed below. Color the bone, using different colors to indicate the various parts. Refer to Figure 9.5 in your textbook for assistance.

surgical neck	lateral epicondyle	olecranon fossa	trochlea
medial epicondyle	coronoid fossa	anatomical neck	
capitulum	deltoid tuberosity	intertubercular sulcus	
lesser tubercle	greater tubercle		

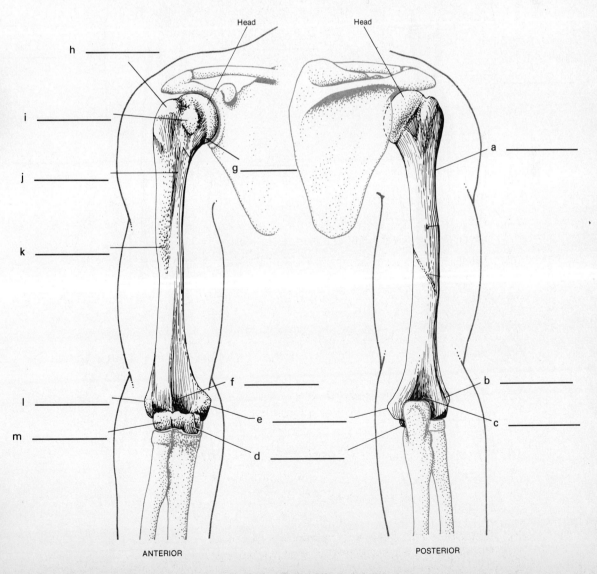

ANTERIOR   POSTERIOR

Figure 9.3

40. Label Figure 9.4, the ulna and radius, using the terms listed below. Color each bone, using different colors for each portion. Refer to Figure 9.6 in your textbook for assistance.

styloid process	radius	head	trochlear notch
olecranon process	coronoid process	ulna	head
tuberosity of radius	neck		

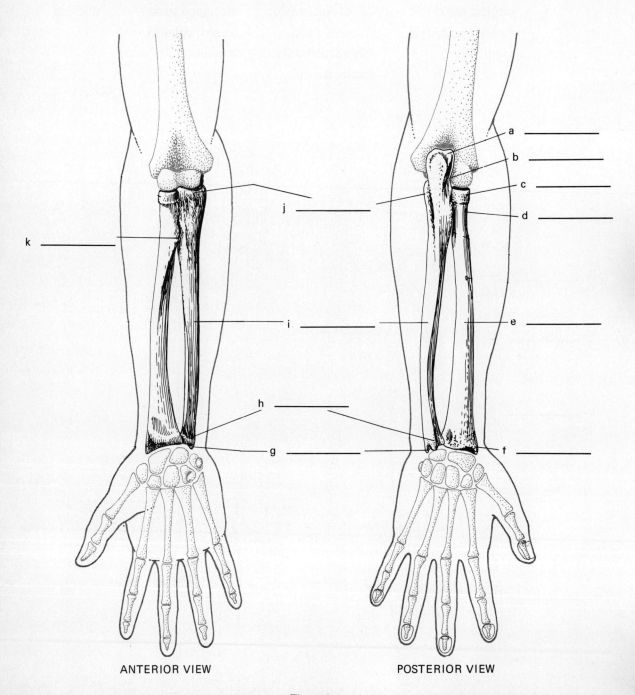

ANTERIOR VIEW   POSTERIOR VIEW

Figure 9.4

41. Label Figure 9.5, the hand, using the terms listed below. Color each bone. Be sure to use different colors for each of the carpal bones. Use the same colors for all of the metacarpals and phalanges. Refer to Figure 9.8 in your textbook for assistance.

phalanges	metacarpals	trapezium	trapezoid	capitate
scaphoid	radius	hamate	triquetrum	pisiform
ulna	lunate			

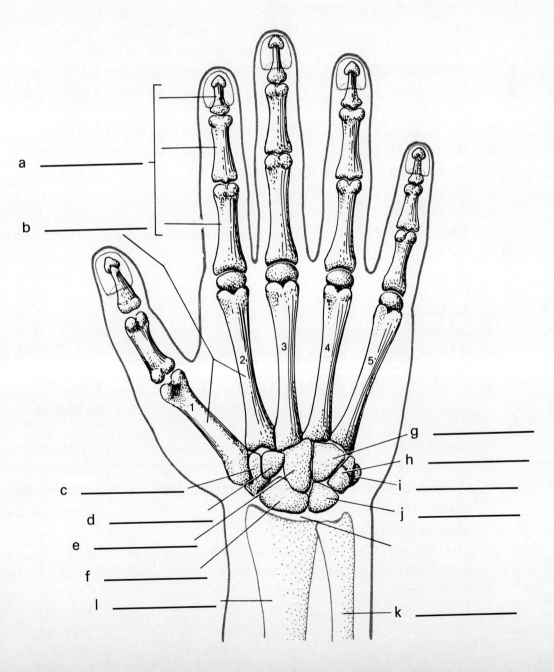

Figure 9.5

42. Each of the coxae, or hip bones, are formed by the fusion of the _____, the _____, and the _____.

43. Ventrally, the two coxae join together at a _____ joint.

44. The cup-like depression with which the head of the femur articulates is known as the _____.

45. When you place your hands on your hips you are placing them on the _____ region of each coxa.

46. When you sit down, you are actually sitting upon the _____ region of each coxa.

47. The pubic symphysis is an _____ articulation.

48. The pubic and ischial rami encircle the _____ foramen.

49. The pelvis contains the coxae, the _____, and the _____.

50. The pelvic cavity is enclosed by the _____ pelvis.

51. The _____ is the bone of the upper leg.

52. The _____ of the femur articulates with the coxae.

53. The greater and lesser _____ serve as attachment points for muscles.

54. The trochanter _____ marks the distal limits of the articular capsule.

55. The _____ surface of the femur is what the kneecap glides.

56. The hip joint is a _____ synovial joint.

57. The two bones of the lower leg are the _____ and _____.

58. The femur articulates distally with the _____.

59. The head of the fibula articulates with the _____ exclusively.

60. The distal lateral _____ of the fibula provides lateral stability to the ankle.

61. The medial _____ of the tibia stabilizes the medial aspects of the ankle.

62. The knee joint is a _____ joint.

63. Structurally, the knee joint resembles _____ different joints.

64. A pair of fibrocartilage pads, the medial and lateral _____, lie between the femoral and tibial surfaces.

65. Label Figure 9.6, the coxa, using the terms listed below. Color the bone, using different colors for the three major regions. Refer to Figure 9.9 of your textbook for assistance.

crest of ilium	anterior superior spine	acetabulum
acetabular notch	superior ramus of pubis	pubic crest
obturator foramen	inferior ramus of pubis	ramus of ischium
tuberosity of ischium	ischial spine	greater sciatic notch
posterior superior spine		

66. Label Figure 9.7, the femur, using the terms listed below. Color the bone, using different colors to highlight the different parts. Refer to Figure 9.11 in your textbook for assistance.

greater trochanter	lesser trochanter	gluteal tuberosity
linea aspera	intercondylar notch	medial condyle
head	patellar surface	lateral condyle
neck		

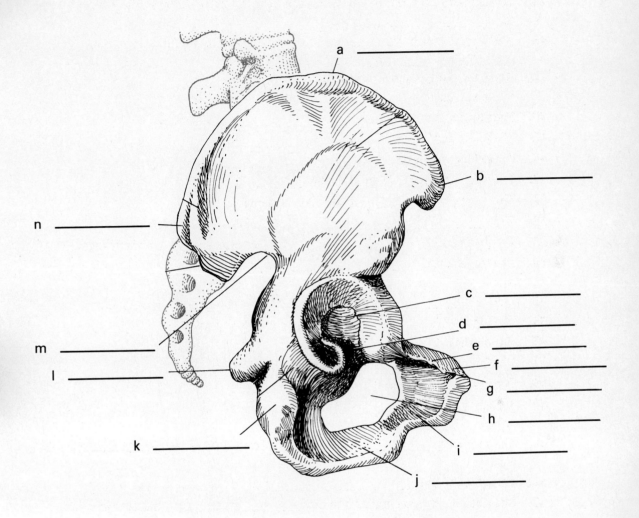

a _____

b _____

n _____

c _____

d _____

e _____

f _____

g _____

h _____

i _____

m _____

l _____

k _____

j _____

Figure 9.6

67. Label Figure 9.8, the tibia and fibula, using the terms listed below. Color the bones. Refer to Figure 9.13 in your textbook for assistance.

| lateral condyle | medial condyle | tibial tuberosity | fibula |
| lateral malleolus | medial malleolus | anterior crest | tibia |

68. There are _____ major ligaments that stabilize the knee joint.

69. The two ligaments inside the joint capsule of the knee joint are the anterior and posterior _____ ligaments.

70. Label Figure 9.9, the knee joint, using the terms listed below. Color each of the structures. Refer to Figure 9.14 in your textbook for assistance.

lateral condyle   posterior cruciate   fibula   fibular

Label: greater trochanter   lesser trochanter   gluteal tuberosity   linea aspera   intercondylar notch
medial condyle   head   patellar surface   lateral condyle   neck

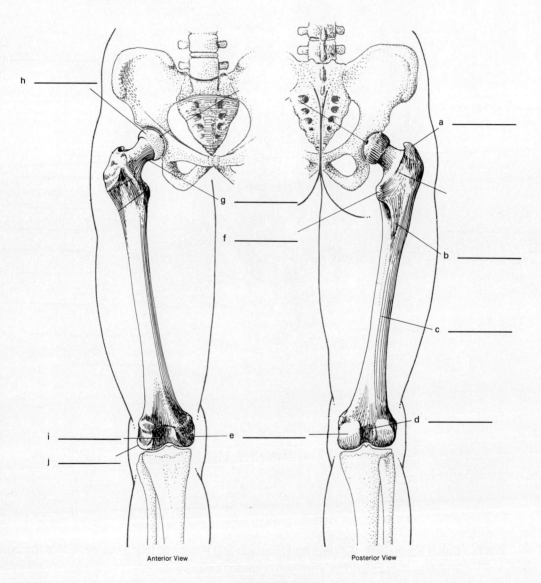

Anterior View   Posterior View

Figure 9.7

collateral ligament	patellar surface	medial condyle
tibial collateral ligament	medial meniscus	anterior cruciate

71. Of the seven bones that form the ankle, only the _____ articulates with the tibia and fibula.

72. The _____ is the "heel bone."

Label: lateral condyle   medial condyle   tibial tuberosity   fibula   lateral malleolus   medial malleolus
         anterior crest   tibia

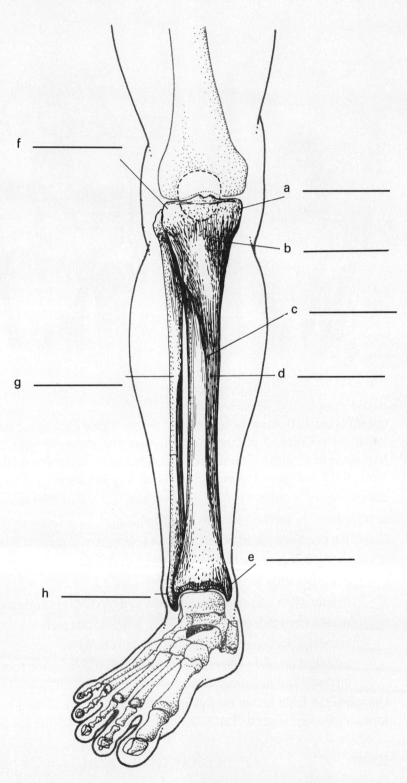

f _____

a _____

b _____

c _____

g _____   d _____

e _____

h _____

Figure 9.8

73.   The sole of the foot is composed of the _____ bones.

74.   Each toe has three phalanges except for the big toe which has _____.

75.   The two arches of the foot are the _____ and the _____.

Label: lateral condyle   posterior cruciate   fibula   fibular collateral ligament   patellar surface
              medial condyle   tibial collateral ligament   medial meniscus   anterior cruciate

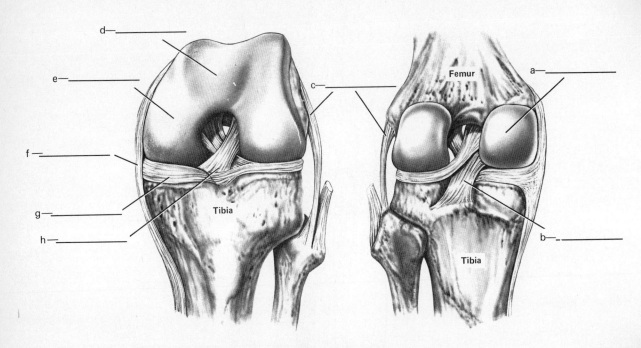

Figure 9.9

76. Label Figure 9.10 using the terms listed below. Color each of the tarsal bones a different color. Color all of the metatarsals and phalanges the same color. Once you have done this compare it with the hand. The similarities of structure should be obvious. Refer to Figure 9.15 in your textbook for assistance.

calcaneous   talus   navicular   cuneiforms   metatarsals
first phalanx   second phalanx   third phalanx

77. Match the condition listed below with the descriptive statement that follows.
    a. sprain   b. dancer's fracture   c. flat feet   d. clubfoot   e. clawfeet
    _____ due to fallen arches
    _____ occurs when a ligament is stretched to the point that collagen fibers tear
    _____ results from an exaggerated median longitudinal arch
    _____ proximal portion of the fifth metatarsal is broken
    _____ inherited problem where abnormal growth results in the foot turning
              medially and inverting

78. Complete the table below by writing in the type of articulation each of the listed joints represents. Refer to Tables 9.1 and 9.2 in your textbook for assistance.

JOINT	TYPE OF ARTICULATION
sternoclavicular	_____
scapulohumeral	_____
olecranal	_____
proximal radioulnar	_____

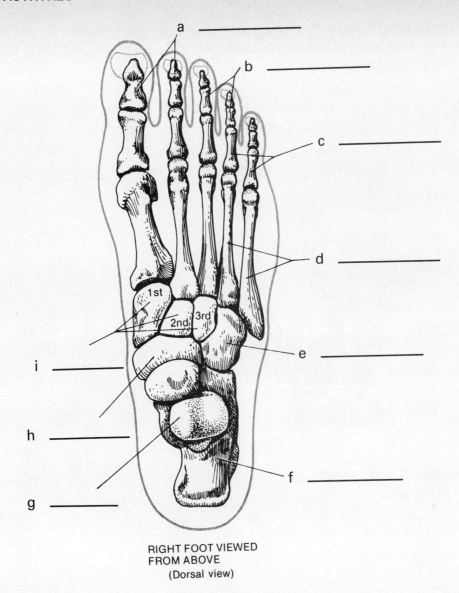

a _____

b _____

c _____

d _____

e _____

i _____

h _____

f _____

g _____

1st

2nd 3rd

RIGHT FOOT VIEWED
FROM ABOVE
(Dorsal view)

Figure 9.10

radiocarpal _____

metacarpophalangeal _____

interphalangeal _____

sacroiliac _____

symphysis pubis _____

hip _____

ankle _____

intertarsal _____

tarsometatarsal _____

knee _____

metatarsophalangeal _____

tibiofibular (proximal) _____

79. Label Figure 9.11, the complete human skeleton, using the terms listed below. Color each of the bones that you label. Refer to the various figures in both chapters 8 and 9 of your textbook.

ribs	sacrum	coccyx	femur	fibula
skull	cervical vertebra	scapula	ulna	calcaneus
humerus	thoracic vertebra	radius	carpals	metacarpals
patella	tibia	clavicle	sternum	ilium phalanges

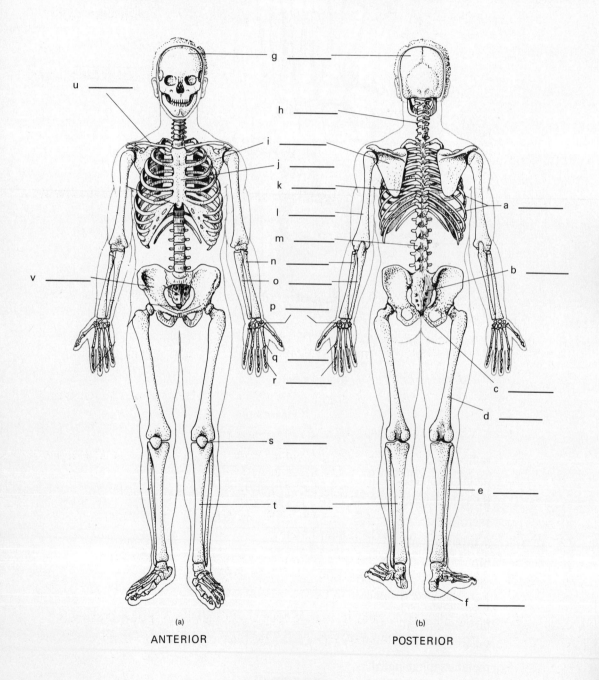

(a) ANTERIOR   (b) POSTERIOR

Figure 9.11

SELF TEST

Circle the correct answer to each question.

1. The pectoral girdle is composed of the scapula and the
   a. clavicle.   b. humerus.   c. ulna.   d. sacrum.   e. radius.

2. The scapulohumeral joint is a _____ joint.
   a. gliding   b. monaxial   c. ball and socket   d. biaxial   e. diarthrosis

3. The depression where the humerus articulates with the scapula is the
   a. obturator foramen.   b. acetabulum.   c. glenoid fossa.   d. olecranon fossa.
   e. coronoid fossa.

4. The region of the humerus which frequently fractures is the
   a. deltoid tuberosity.   b. surgical neck.   c. anatomical neck.   d. greater tubercle.
   e. nutrient foramen.

5. The joint between the metacarpals and the phalanges is a _____
   joint.
   a. ball and socket   b. saddle   c. elipsoid   d. gliding   e. monaxial

6. The two coxae bones join at the
   a. ischium.   b. iliac crest.   c. symphysis pubis.   d. sacroiliac.   e. none of the above

7. The head of the femur fits into the
   a. obturator foramen.   b. glenoid fossa.   c. acetabulum.   d. coronoid fossa.
   e. supraspinatus fossa.

8. The sacroiliac joint is a _____ joint.
   a. ball and socket   b. gliding diarthrosis   c. gliding amphiarthrosis   d. hinge
   e. pivot

9. The distal end of the femur articulates with the
   a. tibia.   b. fibula.   c. tibia and fibula.   d. acetabulum.   e. calcaneous.

10. A ligament from the patella attaches at the
    a. medial malleolus.   b. tibial tuberosity.   c. lateral malleolus.   d. iliac crest.
    e. greater trochanter.

11. The internal ligaments of the knee joint are the
    a. tibial collateral ligaments.   b. fibular collateral ligaments.   c. popiteal ligaments.
    d. cruciate ligaments.   e. patellar ligaments.

12. The tarsal bone that functions as the "heel bone" is the
    a. talus.   b. calcaneous.   c. cuneiform.   d. navicular.   e. cuboid.

13. Distally, the tibia articulates with the
    a. talus.   b. calcaneous.   c. cuneiform.   d. navicular.   e. cuboid.

14. The first or "big" toe contains _____ phalanges.
    a. 1   b. 2   c. 3   d. 4   e. 5

15. A condition that results from an exaggerated longitudinal arch in the foot is
    a. flat feet.   b. dancer's fracture.   c. clawfoot.   d. clubfoot.   e. none of the above

CHAPTER 10
Muscle Tissue

OVERVIEW

The three previous chapters have dealt with support of the body. This chapter begins a consideration of body movement. Movement of the body is powered by muscle. The skeleton and joints act as levers and fulcrums, but it is the skeletal muscles that power them. This chapter introduces you to the structure of muscle and its functioning. You will begin by examining the gross structure of a skeletal muscle. This will then be followed by a look at the mechanism of muscle contraction. Once you have mastered the details of contraction, you will consider the neural control of muscle contraction. It is important to understand that skeletal muscle does not contract without a signal from the nervous system. There then follows a series of discussions on the various aspects of muscle physiology and energetics. Finally there is a discussion of smooth muscle, the muscle of the internal organs.

Muscle is one of the two excitable tissues: nervous tissue is the other one. When examining the basic properties of muscle, especially as relates to the changes in the electrical characteristics of the membrane, it is important to realize that many of these properties also apply to nervous tissue, and will be discussed in that context later in your textbook.

CHAPTER OUTLINE

        c. Excitable membranes and action potentials
        d. The physiological basis for muscular contraction
           (1) The sarcoplasmic reticulum and calcium ions
        e. The control of skeletal muscle activity
    4. Muscular performance
        a. The twitch and tension development
           (1) The effects of varying sarcomere length
        b. The effect of repeated stimulations
        c. Muscle tone
    5. Energetics of muscular activity
        a. ATP and creatine phosphate
        b. ATP and glycogen reserves
           (1) Muscle fatigue
           (2) The recovery period
    6. Specialized skeletal muscle cells
        a. Fast and slow muscle fibers
    7. Muscular performance and endurance
        a. Anaerobic endurance and hypertrophy
        b. Aerobic endurance and the anaerobic threshold
        c. Variations in individual performance
    8. Skeletal muscles and thermoregulation
  C. Cardiac muscle and smooth muscle
    1. Cardiac muscle tissue
    2. Smooth muscle tissue
  D. Clinical patterns
    1. Injuries caused by mechanical forces or pathogens
    2. Conditions characterized by abnormal contraction
    3. Atrophy of the muscle fibers
    4. Conditions caused by muscle fiber changes in response to abnormal motor neuron activity

LEARNING ACTIVITIES

Complete the following items by supplying the appropriate word or phrase.

1. The three distinctive properties of muscle tissue are _____, _____, and _____.

2. List seven functions that muscle tissues carry out in the body.

    _____

    _____

    _____

    _____

    _____

    _____

    _____

3. _____ muscles are the functional units of the muscular system.

4. The collagenous membrane that surrounds each muscle is the _____.

5. The skeletal muscle is broken up into compartments by the _____.

6. A bundle of muscle cells is termed a _____.

7. Each skeletal muscle cell is surrounded by a layer of connective tissue known as the _____.

8. The epimysium of a muscle is continuous with the _____ which connects the muscle to a bone.

9. Match the terms listed below with the appropriate descriptive statements which follow.

    a. sarcolemma   b. sarcoplasmic reticulum   c. myofibrils

    d. myofilaments   e. sarcomere   f. thin filament   g. thick filament

    h. Z line   i. A band   j. I band   k. H zone   l. T tubules

    m. cisternae

    _____ myofilament which is composed primarily of actin

    _____ endoplasmic reticulum of the muscle cell

    _____ invaginations of the muscle cell membrane that contact the cisternae

    _____ expansions of the sarcoplasmic reticulum

    _____ myofilament which is composed of myosin

    _____ functional unit of the myofibril

    _____ cylindrical structures within the muscle cell that extend the length of the cell

    _____ name given the cell membrane of a muscle cell

    _____ microfilaments that make up the sarcomeres

    _____ zones that contain only thin filaments

    _____ dense regions on either end of the sarcomere

    _____ region of the sarcomere that is dominated by thick filaments

    _____ represents the zone of overlap between thin and thick filaments in the sarcomere

10. A typical myofibril will have _____ sarcomeres.

11. Because of their extreme size, skeletal muscle cells are frequently called _____.

12. Because the sarcomeres of adjacent myofibrils are aligned, the entire muscle cell has a _____ appearance under a light microscope.

13. Label figure 10.1, structure of a skeletal muscle, using the terms listed below. Color the various structures.

    Epimysium   Perimysium   Fasciculus   Endomysium   Fiber
    Myofibril

14. The thick filaments are composed of _____.

15. In addition to actin, thin filaments contain the proteins _____ and _____.

16. Thin filaments have _____ sites that can attach to cross-bridges from the thick filaments.

17. The repeated pivoting of cross-bridges causes the thin filaments to slide _____.

18. The inward movement of the thin filaments causes the sarcomere to _____.

19. The mechanism of muscle shortening is known as the _____ theory of contraction.

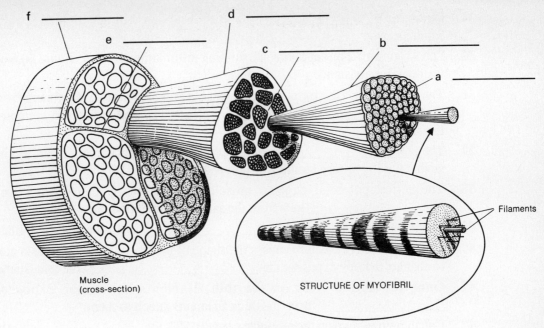

f _____
e _____
d _____
c _____
b _____
a _____

Filaments

Muscle
(cross-section)

STRUCTURE OF MYOFIBRIL

Figure 10.1

20. The extracellular fluid is high in _____ ions while the cytoplasm (intracellular fluid) is high in _____ ions.

21. The resting membrane condition has an excess of _____ charge on the outside of the cell membrane and an excess of _____ charges on the inside of the membrane.

22. Whenever electrical charges are separated, a _____ difference exists.

23. The potential difference that exists across the cell membrane due to the separation of charges by the membrane is termed the _____ potential.

24. The unit of measurement for potential difference is the _____.

25. The transmembrane potential for a typical skeletal muscle cell is _____ mV.

26. Chemical stimuli which open the sodium channels increase the influx of positive charges into the cell and cause _____ of the membrane.

27. Chemical stimuli which open the potassium channels increase the loss of positive charges from the inside of the cell and cause _____ of the membrane.

28. Muscle and nerve cell differ from other cells in that they possess the property of _____.

29. The conducted change in the membrane potential that can occur on muscle and nerve cells is known as the _____ potential.

30. The action potential is initiated once the membrane depolarizes to the _____ level.

31. During the first phase of the action potential, depolarization, _____ ions are flowing into the cell.

32. During the second phase of the action potential, repolarization, _____ ions are flowing out of the cell.

33. At threshold, the membrane characteristics change so that the membrane suddenly becomes highly permeable to _____.

34. During the _____ an excitable cell will not respond to a liminal stimulus.

35. Label figure 10.2 by supplying the missing information. Refer to figure 10.6 of your textbook for assistance.

36. The action potential is propagated from one segment of the cell membrane to another by local _____ which develop and bring the membrane to threshold.

37. The _____ principle states that all liminal stimuli, regardless of their strength, will produce identical action potentials.

38. The action potential triggers muscle contraction by causing a release of _____ ions from the sarcoplasmic reticulum.

39. Calcium ions interact with the _____ molecule of the thin filaments.

40. The interaction of calcium causes the shape of the troponin- tropomyosin complex to change, thereby exposing the _____ sites of the thin filaments.

41. Once the active sites on the thin filaments have been exposed, the _____ from the thick filaments attach to them.

42. Calcium gives myosin the ability to convert ATP to _____, thereby providing the energy for contraction.

43. Once the contraction is terminated, calcium is pumped back into the _____ reticulum.

44. The removal of calcium from the sarcomeres permits the troponin- _____ complex to reform, thereby covering the active sites again.

45. All of the muscle cells controlled by a single motor neuron constitutes a _____ unit.

46. The motor neuron releases the chemical _____ into the gap between it and the muscle cell.

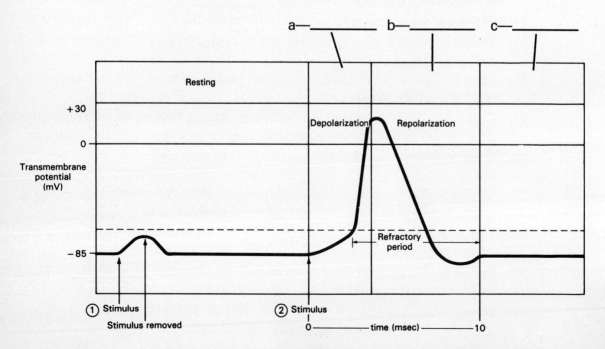

Figure 10.2

47. Acetylcholine attaches to receptor sites on the motor end plate of the sarcolemma and initiates an _____.

48. Acetylcholine is immediately broken down by the enzyme _____.

49. Label all of the structures in figure 10.3, the myoneural junction, using the terms listed below. Color each structure. Refer to figure 10.7 of your textbook for assistance.

    muscle cell   ACh receptors   synaptic cleft   synaptic vesicle
    neuron

50. Contraction where tension on the muscle develops and shortening occurs is termed _____.

51. Contraction where tension on the muscle develops but no shortening occurs is termed _____.

52. Postural muscles which maintain the bodie's position utilize _____ contractions, while muscles involved in movement utilize _____ type contraction.

53. A muscle _____ represents a single stimulus/contraction/relaxation sequence.

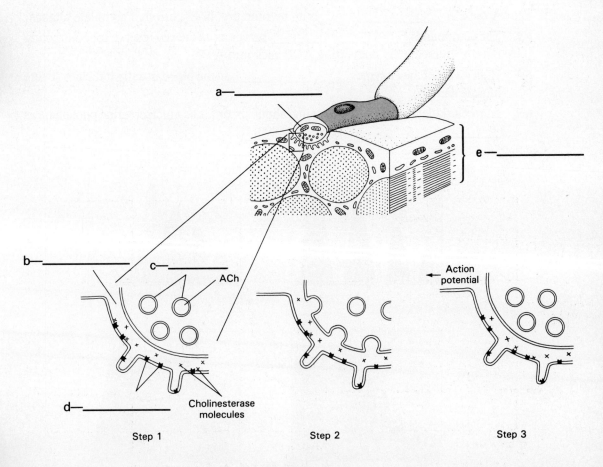

Figure 10.3

54. For the the various phases of a muscle twitch which are listed below, write in the approximate time that each takes.

    _____ latent period

    _____ contraction phase

    _____ relaxation phase

55. The force which a muscle can exert varies with its relaxed _____.

56. Within the optimal range of sarcomere lengths, the maximum number of _____ can form, and tension production is highest.

57. Label each graph in figure 10.4 with the correct physiological term from those listed below. Refer to figure 10.11 of your textbook for assistance.

    complete tetanus   treppe   incomplete tetanus   summation

58. Elimination of the relaxation period by rapid stimuli results in a sustained tension known as complete _____.

59. The physiological basis for tetanus is that action potentials are arriving so rapidly that the sarcoplasmic reticulum does not have time to reabsorb _____ ions from the sarcomeres.

60. Tension in whole muscles may be increased by recruiting additional _____.

61. Muscle _____ is due to the fact that there are a certain number of motor units active in all muscles, even when they are relaxed.

62. A muscle _____ is an involuntary, brief period of complete tetanus.

63. The compound _____ serves as an energy reserve for contracting muscle, immediately converting ADP back into ATP.

64. High levels of the enzyme _____ in the blood usually indicates severe muscle damage.

65. At rest a skeletal muscle cell contains about six times as much creatine phosphate as _____.

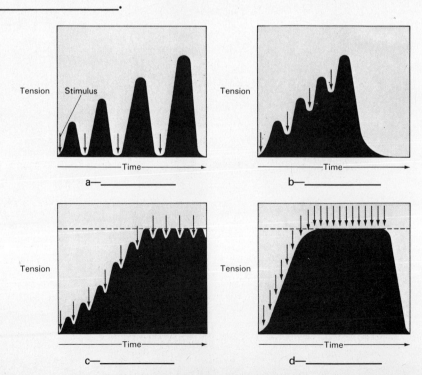

Figure 10.4

66. When the mitochondria cannot generate ATP fast enough to meet the muscle cell's demand, the cell can generate ATP through anaerobic _____.

67. During anaerobic glycolysis, pyruvic acid is converted into _____ acid.

68. During peak levels of exertion, roughly two thirds of the energy requirements of a skeletal muscle cell are met by _____.

69. Muscle fatigue is brought about by depletion of glycogen reserves and increased levels of _____.

70. The leaking of calcium from the sarcoplasmic reticulum into the myofibrils following death results in the formation of cross bridges between thin and thick filaments, a condition termed _____.

71. Following extensive muscle activity, lactic acid is sent to the liver and converted into _____.

72. The glucose produced by the liver from lactic acid is then used by the muscles to synthesize _____.

73. The increased oxygen consumption that is necessary to metabolize lactic acid is known as the _____, and is the reason that breathing levels continue higher than normal even after exercise has ceased.

74. Slow fibers contain the the red pigment _____ which enhances their ability to absorb oxygen.

75. The capillary network is much more extensive around _____ fibers.

76. Muscles rich in slow fibers have a much greater _____ than do those rich in fast fibers.

77. Muscles dominated by slow fibers are known as _____ muscles.

78. The "dark meat" in poultry is muscle that contains a high proportion of _____ fibers.

79. A weight lifter would exercise in such a manner that would increase his or her _____.

80. Athletes that require a lot of strength train in such a manner as to induce _____ of the muscles.

81. Muscles trained for aerobic endurance can use _____ and _____ as energy sources in addition to the usual carbohydrates.

82. Aerobic conditioning improves both _____ and _____ efficiency in addition to that of muscle.

83. Aerobic conditioning requires that energy demands of muscle be maintained at or below the _____ threshold.

84. A person interested in becoming a better tennis player would seek to increase their _____ endurance.

85. A muscle which is not exercised will begin to _____.

86. When muscle contracts there are two periods of heat. The first corresponds to the actual period of exercise and is termed _____, while the second which corresponds to the oxygen debt period is termed the _____ heat.

87. Fine contraction of the skeletal muscle which is aimed at increasing body heat is known as _____.

88. The intercellular connections between cardiac muscle cells are the _____ discs.

89. Unlike skeletal muscle, cardiac muscle cells do not require _____ stimulation.

90. The _____ period for cardiac muscle is much longer than that of skeletal muscle.

91. As a result of the long refractory period partial or complete _____ does not occur in the heart.

92. The property of smooth muscle that permits it to change length without changing tension is termed _____.

93. Smooth muscle which is innervated in motor units similar to skeletal muscle is termed _____.

94. Many _____ smooth muscle fibers lack a direct contact with any motor neurons.

95. Match the clinical condition listed below with the descriptive statement which follows.

    a. trichinosis   b. fibrositis   c. hyperkalemia   d. tetanus
      e. muscular dystrophy   f. myasthenia gravis   g. botulism

    _____ muscle weakness caused by a loss of acetylcholine receptors

    _____ genetic disease that produces progressive deterioration and weakness of the muscles

    _____ bacterial infection that causes spastic contraction of the muscles

    _____ inflammation of the connective tissues surrounding muscle

    _____ caused by the invasion of muscle by parasitic worms

    _____ paralysis produced by bacterial poisoning

    _____ elevated potassium levels in the blood which result in paralysis of muscles

SELF TEST

Circle the correct answer to each question.

1. Which of the following is a function of muscle?
   a. movement   b. maintenance of position   c. regulating blood flow   d. supporting soft tissue   e. more than one of the above is correct.

2. The connective tissue envelope that surrounds skeletal muscle is the
   a. epimysium   b. perimysium.   c. endomysium.   d. fasciulus.   e. tendon.

3. The structural units of the myofibrils are the
   a. fasciculi.   b. sarcomeres.   c. myosin.   d. thick filaments.   e. thin filaments.

4. The areas of the sarcomere which consists of nothing but thin filaments is the
   a. A-band.   b. Z-line.   c. I-band.   d. H-zone   e. M-line.

5. The expanded regions of the sarcoplasmic reticulum are the
   a. t-tubules.   b. cisternae.   c. cross bridges.   d. zone of overlap.   e. active sites.

6. Thin filaments are composed of
   a. actin.   b. troponin.   c. tropomyosin.   d. active sites.   e. more than one of the above is correct.

7. The sarcomeres and the muscle cell contract when

   a. the thick filaments slide outward.   b. the thin filaments slide outward.   c. the thin filaments slide inward.   d. the active sites cover up.   e. the cross bridges break loose.

8. During the depolarization phase of the action potential

   a. sodium is flowing out of the cell.   b. potassium is flowing into the cell.   c. sodium is flowing into the cell.   d. potassium is flowing into the cell.   e. none of the above.

9. The action potential is conducted across the cell membrane by the generation of

   a. refractory periods.   b. subliminal stimulus.   c. local current.   d. repolarization.   e. none of the above.

10. Calcium causes the uncovering of the active sites on the thin filaments by interacting with

    a. tropomyosin.   b. actin.   c. troponin.   d. myosin.   e. none of the above.

11. A given motor unit contains 1000 skeletal muscle cells. It will contain _____ motor neurons.

    a. 1000   b. 100   c. 10   d. 1   e. cannot be determined

12. The action potential is transmitted from the neuron to the motor end plate of the muscle cell by

    a. sparks.   b. acetylcholine.   c. acetylcholinesterase.   d. troponin.   e. calcium.

13. If a skeletal muscle is stretched prior to contraction, the tension developed will be

    a. greater than the unstretched state.   b. less than the unstreched state.   c. greater or lesser depending upon the amount of stretch.   d. the same as the unstretched state.   e. none of the above.

14. If a muscle is stimulated repetitively so that there is no time to relax, the condition produced will be

    a. treppe.   b. summation.   c. incomplete tetanus.   d. complete tetanus.   e. tonus.

15. The immediate source of energy to regenerate ATP in skeletal muscle is

    a. glycogen.   b. glucose.   c. creatine phosphate.   d. fatty acids.   e. amino acids.

CHAPTER 11
The Muscular System

OVERVIEW

This chapter examines muscles as organs interacting as part of a system. It begins by examining the organization of the fibers in the skeletal muscles and then proceeds to a consideration of the functioning of muscles when they are attached to the skeletal system. The mechanism by which muscles make the the skeleton move, namely the principle of the lever is then considered. A brief introduction to the terminology of muscle then follows, and the remainder of the chapter deals with the various skeletal muscles of the body.

The organization of the muscular system follows the pattern set by the skeleton. There is first of all an examination of the major muscles associated with the axial division and then the muscles associated with the appendicular division are considered. There are five important aspects of each skeletal muscle which you should master: these are the muscle name, its origin, insertion, mode of action, and innervation. Much of this information has been placed in tables for you in your textbook and you should make full use of these. Be sure to examine each of the illustrations in your text as well. Finally, try and locate as many of the muscles on your own body as possible. This will make what you read in your textbook much more understandable.

CHAPTER OUTLINE

    A. Introduction
    B. Biomechanics and muscular anatomy
       1. Skeletal muscle fiber organization
       2. Structural interactions
          a. Levers in action
    C. Muscle terminology
       1. Origins, insertions, and actions

LEARNING ACTIVITIES

Complete each of the following items by supplying the appropriate word or phrase.

1. In _____ muscles the fasciculi are parallel to the long axis of the muscle.

2. A skeletal muscle cell can contract effectively until it has shortened _____ percent of its original length.

3. In a _____ muscle the muscle fibers are based over a broad area but come together to a common attachment.

4. In a _____ muscle the fasciculi form a common angle with the tendon.

5. The rectus femoris is an example of a _____ muscle.

6. An example of a multipennate muscle is the _____ of the shoulder.

7. Muscles in which the fiber arrangement is circular are usually referred to as _____ muscles.

8. The balance point of a lever is termed the _____.

9. Levers can be used to change the amount and the _____ of movement.

10. Levers can also be used to increase or decrease the amount of _____ being applied.

11. Levers can change the relationship between _____, _____, and the amount of _____.

12. The attachment point of a muscle which remains stationary or shows very little movement is the _____, while the attachment point that shows considerable movement is the _____.

13. The _____ of a muscle can be thought of as the function which the muscle carries out.

14. Almost all skeletal muscles originate or insert upon the _____.

15. The muscle which is mainly responsible for a particular movement is termed the _____ mover.

16. A _____ aids the prime mover in its action.

17. When two muscles or groups of muscles have actions which are the opposite of one another they are known as _____.

18. Match the muscle listed below with the statement of action which follows. Refer to table 11.2 of your textbook for assistance

    a. buccinator b. depressor labii c. levator labii d. mentalis e. orbicularis oris f. risorius g. zygomaticus h. corrugator supercilii i. levator palpebrae j. orbicularis oculi k. procerus l. nasalis m. auricularis n. frontalis o. occipitalis p. platysma

    _____ compresses the bridge of the nose

    _____ compresses the cheeks

    _____ muscle which depresses the lip

    _____ muscle which raises the upper lip

    _____ responsible for closing the eye

    _____ causes a change in the shape of the nostrils

    _____ tenses the skin of the neck and depresses the mandible

    _____ muscle that draws the corners of the mouth back and up

    _____ muscle which raises the upper eyelid

    _____ moves the external ear

    _____ muscle which elevates and protrudes the lower lip

    _____ muscle that causes the forehead to wrinkle

    _____ muscle which tenses and retracts the scalp

    _____ pulls the skin down and forward, thereby wrinkling the brow

    _____ draws the corner of the mouth to the side

19. Most of the muscles of facial expression are innervated by the _____ nerve.

20. Label figure 11.1, facial muscles of expression, using the terms listed below. Refer to figure 11.5 of your textbook for assistance. Color each the muscles which you label.

    occipitalis   auricular   frontalis   orbicularis oculi
    nasalis   zygomaticus   orbicularis oris   risorius
    buccinator platysma

21. The major muscles of mastication are the _____, _____, and _____.

22. Label figure 11.2, the extrinsic muscles of the eye, using the terms listed below. Color each of the muscles labeled. Refer to figure 11.6 of your textbook for assistance.

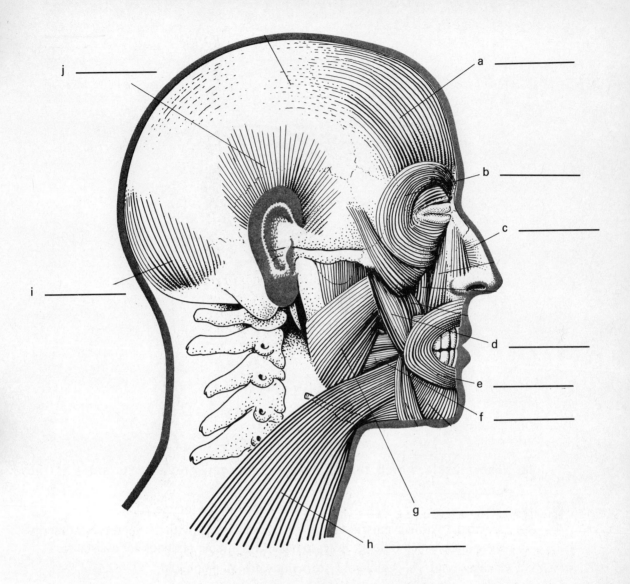

j _____

a _____

b _____

c _____

i _____

d _____

e _____

f _____

g _____

h _____

Figure 11.1

superior rectus   inferior rectus   medial rectus   lateral
rectus   superior oblique   inferior oblique

23. The inferior rectus, medial rectus, superior rectus, and inferior oblique are all inner-
    vated by the _____ nerve.

24. The lateral rectus is innervated by the _____ nerve.

25. The superior oblique is innervated by the _____ nerve.

26. The tongue muscle which originates on the medial surface of the mandible around
    the chin is the _____.

27. The muscle which elevates the tongue and depresses the fleshy palate is the
    _____.

Label: superior rectus   inferior rectus   medical rectus   lateral rectus   superior oblique   inferior oblique

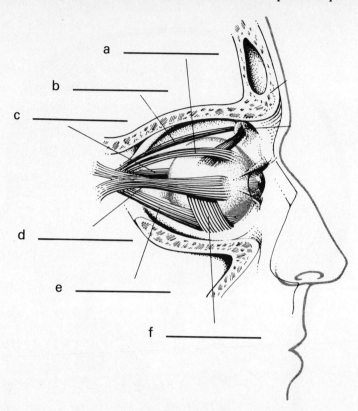

a _____

b _____

c _____

d _____

e _____

f _____

Figure 11.2

28. The muscle which retracts the tongue and elevates its sides is the _____.

29. The _____ depresses and retracts the tongue.

30. Match the extrinsic muscles of the larynx listed below with the appropriate descriptive phrase which follows. Refer to table 11.6 of your textbook for assistance.

a. digastricus   b. geniohyoid   c. mylohyoid   d. omohyoid

e. sternohyoid   f. sternothyroid   g. stylohyoid   h. thyrohyoid

_____ originates on the clavicle and inserts on the hyoid bone

_____ elevates the thyroid gland

_____ originates on the hyoid bone

_____ originates on the styloid process of the temporal bone

_____ inserts on the thyroid cartilage of the larynx

_____ elevates the floor of the mouth and/or depresses the mandible

_____ one of its two bellies inserts on the hyoid bone

31. The spinalis group of superficial spinal muscles includes the _____, _____, and _____.

32. In the longissimus group of spinal muscles there are three muscles, the longissimus _____, _____, and _____.

33. The deep spinal extensors consist of the _____ group.

34. The major spinal flexors are the _____, _____, and _____.

35. Match the muscles listed below with the appropriate descriptive phrase that follows.

a. scalenes   b. external intercostals   c. internal intercostals
d. transversus abdominus   e. diaphragm   f. rectus abdominus
g. internal oblique   h. transverse thoracis   i. external oblique

_____ innervated by the cervical spinal nerves

_____ contraction of this muscle expands the thoracic cavity

_____ originate on the lumbodorsal fascia and iliac crest

_____ originate at the superior border of each rib and depresses the ribs

_____ depress the ribs and flexes the vertebral column

_____ originate on the cartilage of the lower ribs, iliac crest, and lumbodorsal fascia

_____ originate on the inferior border of each rib

_____ insert on the cartilage of the ribs and depress the ribs

_____ originate on the lower eight ribs and insert on the linea alba and iliac crest

36. The two most common types of hernias are the _____ and the _____.

37. The pelvic floor is termed the _____.

38. The perineum is divided into two triangles. The anterior triangle is the _____.

39. The muscle that stiffens the penis and ejects urine or semen is the _____.

40. The majority of the muscles of the pelvic floor are innervated by the _____ nerve.

41. The muscles that make up the urogenital triangle are the _____ and _____.

42. The pelvic diaphragm is composed of the _____, _____, _____, and the _____.

43. Match the muscle listed below with the appropriate descriptive phrase that follows. Refer to table 11.10 of your textbook for assistance.

a. levator scapulae   b. pectoralis minor   c. rhomboideus major   d. serratus anterior
e. sternocleidomastoid   f. subclavius   g. trapezius

_____ depresses and protracts the shoulder

_____ adducts and rotates the scapula laterally

_____ innervated by the long thoracic nerve

_____ inserts on the mastoid region of the skull

_____ inserts on both the clavicle and scapula

_____ depresses and protracts the shoulder

_____ elevates the scapula

44. The three muscles of the upper arm that are innervated by the suprascapular nerve are the _____, _____, and the _____

45. The _____ inserts on the deltoid tuberosity of the humerus.

46. The _____ adducts and flexes the humerus.

47. The origin of the teres major is the inferior angle of the _____.

48. The muscle which flexes, adducts, and medially rotates the humerus is the _____.

49. The _____ is the muscle of the upper arm which is innervated by the thoracodorsal nerve.

50. Label figure 11.3, muscles that move the shoulder, using the terms listed below. Color each of the muscles. Refer to figure 11.14 of your textbook for assistance.

    a. serratus anterior   b. trapezius   c. pectoralis minor   d. rhomboid minor
    e. rhomboid major   f. levator scapulae

51. Write in the name of the muscle that best fits the description given. Refer to table 11.12 of your textbook for assistance.

    _____ supinates the forearm

    _____ receives innervation from the ulnar nerve

    _____ inserts on the radial tuberosity

    _____ extends and abducts the palm

    _____ inserts on the palmar aponeurosis

    _____ has three heads of origin

    _____ inserts on the ulnar tuberosity

    _____ extends and adducts the palm

    _____ inserts on the base of the second metacarpal

    _____ originates on the lateral epicondyle of the humerus and
                          inserts on the styloid process of the radius

    _____ pronates the forearm

52. The _____ adducts the thumb.

53. The _____ flex the fingers.

54. The _____ abducts the thumb.

55. The fingers and palm are extended by the _____.

56. Label figure 11.4, muscles of the upper arm, with the terms listed below. Color each of the muscles. Refer to figure 11.15 in your textbook for assistance.

    coracobrachialis   triceps (long head)   triceps (lateral head)

    biceps (short head)   biceps (long head)   supraspinatus

    infraspinatus   teres minor   teres major

57. Label figure 11.5, the forearm, with the terms listed below. Color each of the muscles. Refer to figures 11.16, 11.17 and 11.18 in your textbook for assistance.

    extensor carpi radialis   brachioradialis   extensor digitorum

    abductor pollicis longus   extensor carpi ulnaris   supinator

    flexor digitorum   flexor carpi ulnaris

58. The three muscles of the gluteal group are the gluteus _____, _____, and _____.

59. The two muscles of the group responsible for lateral rotation of the thigh are the _____ and _____.

60. The adductor brevis, longus, and magnus originate from the _____ region of the coxae.

61. The pectineus muscle flexes and _____ the thigh.

62. The iliopsoas is innervated by the _____ nerve.

63. Write the name of the muscle that best fits the descriptive statements below. Refer to table 11.15 in your textbook for assistance.

    _____ lateral member of the quadriceps group

    _____ inserts on the head of the fibula and the lateral condyle of
                          the tibia

    _____ originates on the proximal shaft of the tibia

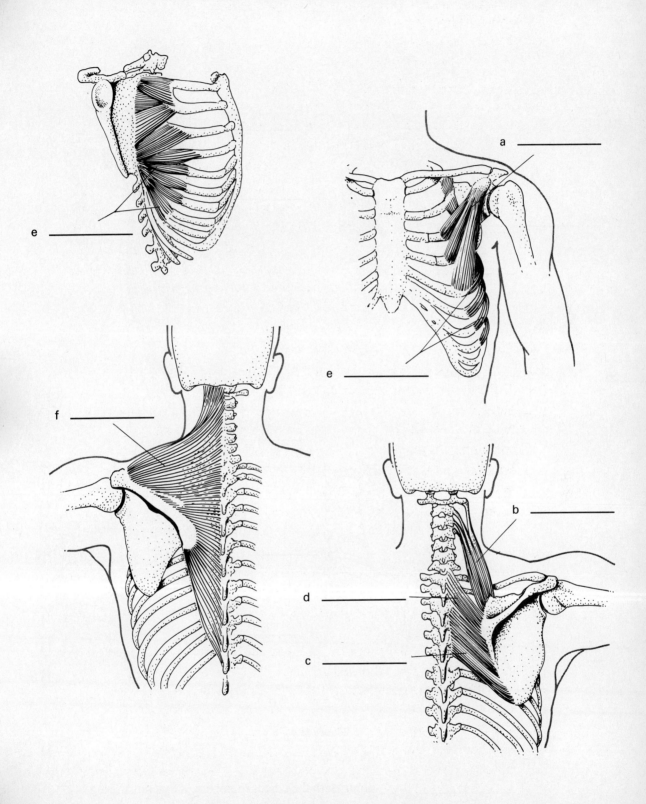

Figure 11.3

Label: coracobrachialis   triceps (long head)   triceps (lateral head)   biceps (short head)   biceps (long head)
supraspinatus   infraspinatus   teres minor   teres major

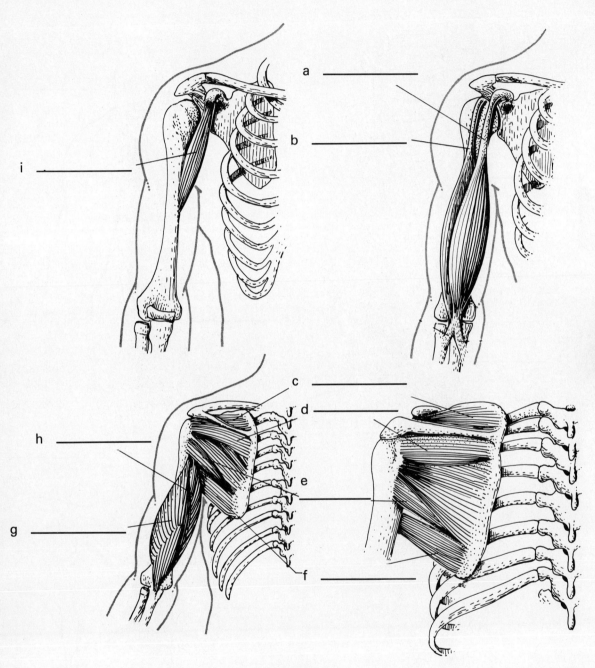

a _____

b _____

i _____

c _____

d _____

h _____

e _____

g _____

f _____

Figure 11.4

_____ along with the vastus members, it extends the leg

_____ flexes the leg and thigh, and laterally rotates the thigh

_____ along with the semitendinosus flexes the leg, extends,
adducts, and medially rotates the thigh

_____ innervated by the obturator nerve

Label: extensor carpi radialis   brachioradialis   extensor digitorum   abductor pollicis longus
extensor carpi ulnaris   supinator flexor digitorum   flexor carpi ulnaris

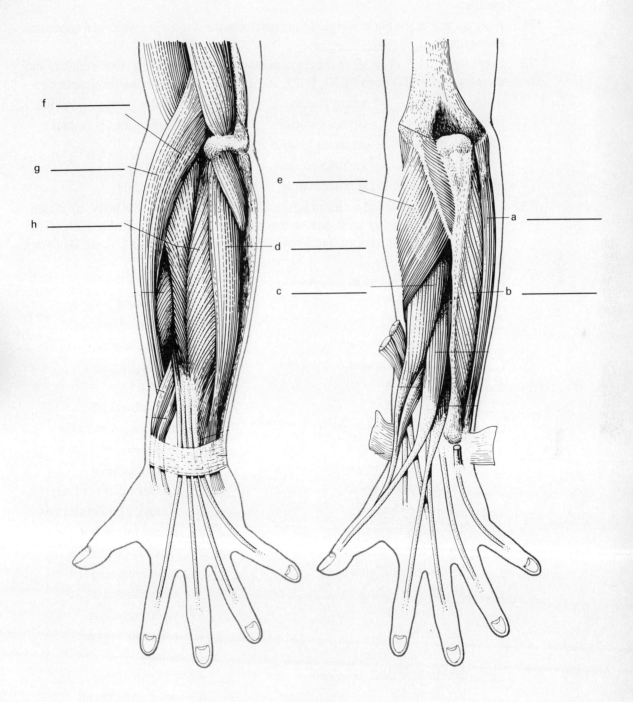

Figure 11.5

64. The _____ dorsiflexes the foot.

65. The _____ shares a common tendon with the soleus and inserts on the calcaneous.

66. The plantar toe flexor muscles are the _____.

67. Dorsiflexion of the toes is caused by the _____.

68. Eversion of the foot is caused by the _____ muscles.

69. The tibialis posterior adducts and _____ the foot.

70. Tennis elbow is an inflammation at the origin of the _____ muscles.

71. A partial tear at the lateral origin of the gastrocnemius muscle results in a condition known as _____.

72. Label figure 11.6, muscles of the leg using the terms listed below. Color each of the muscles. Refer to figures 11.20, 11.21, and 11.22 in your textbook for assistance.

tensor fascia lata	rectus femoris	vastus lateralis	sartorius
peroneus longus	tibialis anterior	extensor digitorum	soleus
adductor magnus	adductor longus	gracilis	
vastus medialis	gastrocnemius	gluteus maximus	
biceps femoris	semitendinosus	semimembranosus	

73. Identify the numbered muscles in figure 11.7, anterior view of the body, by matching the muscle number with one of the muscles listed below. Color each of the muscles. Refer to the various figures of muscles found throughout chapter 11 of your textbook.

_____ 1. tibialis anterior

_____ 2. sartorius

_____ 3. gracilis

_____ 4. adductor magnus

_____ 5. adductor longus

_____ 6. pectineus

_____ 7. iliopsoas

_____ 8. rectus abdominus

_____ 9. external oblique

_____ 10. serratus anterior

_____ 11. latissimus dorsi

_____ 12. triceps brachii

_____ 13. flexor carpi radialis

_____ 14. extensor radialis

_____ 15. extensor digitorum

_____ 16. biceps brachii

_____ 17. trapezius

_____ 18. sternohyoid

_____ 19. orbicularis oris

_____ 20. levator labii superioris

_____ 21. frontalis

_____ 22. temporalis

_____ 23. masseter

_____ 24. buccinator

_____ 25. sternocleidomastoid

_____ 26. sternothyroid

_____ 27. pectoralis major

_____ 28. deltoid

_____ 29. coracobrachialis

_____ 30. external intercostals

_____ 31. brachialis

_____ 32. brachioradialis

_____ 33. extensor carpi radialis

_____ 34. abductor pollicis longus

_____ 35. tensor fascia lata

_____ 36. quadriceps femoris

_____ 37. rectus femoris

_____ 38. vastus lateralis

_____ 39. vastus medialis

_____ 40. peroneus longus

_____ 41. soleus

_____ 42. gastrocnemius

_____ 43. orbicularis occuli

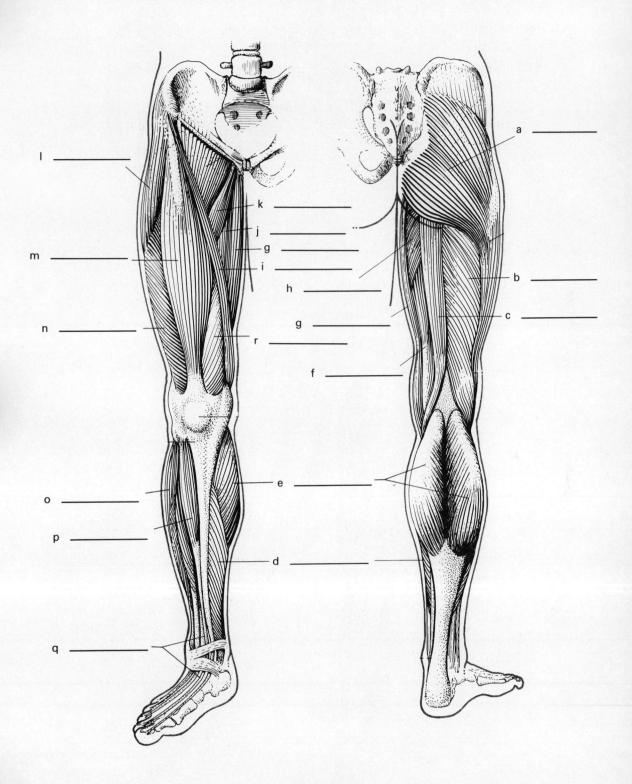

Figure 11.6

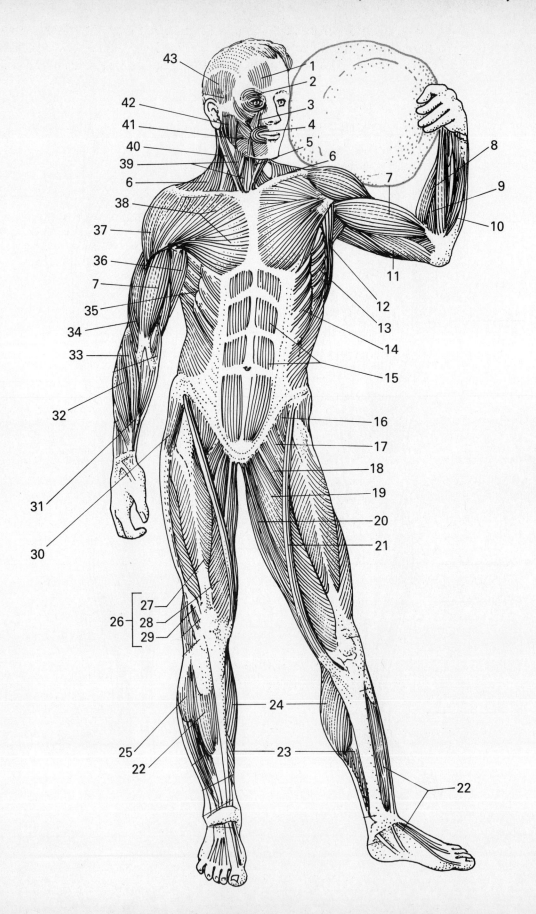

Figure 11.7

74. Identify the numbered muscles in figure 11.8, posterior view of the body, by matching the muscle number with one of the muscles listed below. Color each of the muscles. Refer to the various figures of muscles found throughout chapter 11 of your textbook.

_____ 1. soleus

_____ 2. extensor digitorum longus

_____ 3. peroneus longus

_____ 4. gastrocnemius

_____ 5. plantaris

_____ 6. biceps femoris

_____ 7. vastus lateralis

_____ 8. adductor magnus

_____ 9. tensor fascia lata

_____ 10. gluteus maximus

_____ 11. gluteus medius

_____ 12. external oblique

_____ 13. latissimus dorsi

_____ 14. rhomboid major

_____ 15. teres major

_____ 16. teres minor

_____ 17. brachioradialis

_____ 18. extensor digitorum communis

_____ 19. extensor carpi radialis

_____ 20. semimembranosus

_____ 21. semitendinosus

_____ 22. gracilis

_____ 23. abductor pollicis longus

_____ 24. extensor carpi ulnaris

_____ 25. flexor carpi ulnaris

_____ 26. aconeus

_____ 27. biceps brachii

_____ 28. triceps brachii (lateral head)

_____ 29. triceps brachii (long head)

_____ 30. deltoid

_____ 31. trapezius

_____ 32. sternocleidomastoid

_____ 33. occipitalis

SELF TEST

Circle the correct answer to each question.

1. The muscle pattern in which the fibers are arranged in a feather- like pattern is the
   a. parallel.   b. unipennate.   c. circular.   d. convergent.   e. bipennate.

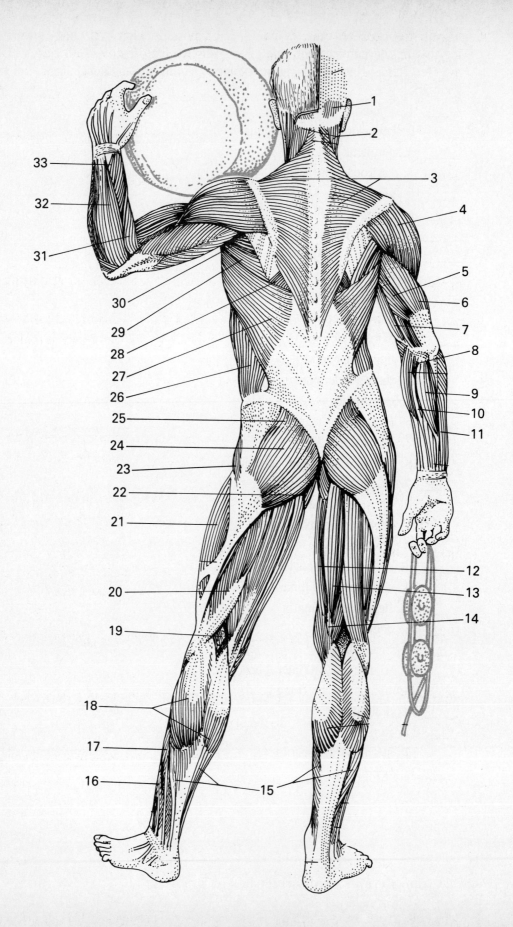

Figure 11.8

2. The balance point of a lever is the
   a. resistance.   b. end.   c. fulcrum.   d. force point.   e. speed increase.

3. The attachment point of a muscle which shows the least movement is its
   a. origin.   b. insertion.   c. action.   d. innervation   e. none of the above.

4. The prime mover for mastication would be the
   a. orbicularis oris.   b. buccinator   c. zygomaticus.   d. masseter.   e. temporalis.

5. The muscles of the tongue are all innervated by the _____ nerve.
   a. trigeminal   b. facial   c. vagus   d. mental   e. hypoglossal

6. Which of the following is an oblique muscle?
   a. internal intercostal   b. external intercostal   c. transversus   d. scalenes   e. more than one of the above is correct.

7. The thoracic and abdominopelvic cavities are separated by the
   a. external intercostals.   b. internal intercostals.   c. transversus.   d. diaphragm.
   e. scalenes.

8. Which of the following muscles is not associated with movement of the shoulder?
   a. trapezius   b. sternocleidomastoid   c. serratus anterior   d. pectoralis minor
   e. biceps brachii

9. The most important abductor of the upper arm is the
   a. biceps brachii.   b. triceps brachii.   c. deltoid.   d. pectoralis major.
   e. brachialis.

10. The muscle which is principally responsible for extension of the forearm is the
    a. biceps brachii.   b. triceps brachii.   c. flexor carpi.   d. brachioradialis.
    e. ulnaris.

11. The ileotibial tract is formed by the insertions of the tensor fascia lata and the
    a. gluteus maximus.   b. gluteus minimus.   c. gluteus medius.   d. piriformis.
    e. gastrocnemius.

12. The hamstring muscles include the
    a. biceps femoris.   b. rectus femoris.   c. platysma.   d. soleus.   e. more than one
    of the above is correct.

13. Extension of the leg is the function of the
    a. vastus lateralis.   b. vastus medialis.   c. rectus femoris.   d. vastus intermedius.
    e. more than one of the above is correct.

14. The gastrocnemius is responsible for
    a. palmar flexion.   b. palmar extension.   c. plantar flexion.   d. plantar extension.
    e. plantar eversion.

15. Dorsiflexion of the foot is a function of the
    a. sartorius.   b. gracilis.   c. biceps femoris.   d. semitendinosus.   e. tibialis
    anterior.

CHAPTER 12

The Nervous System:
Neural Tissue

OVERVIEW

All of our abilities are ultimately limited to the contraction of muscles. It is the order in which we contract our muscles that we call human behavior. The order of these events is determined and controlled by the nervous system, which is therefore the center of all human behavior.

This chapter is the first of six which deal with various aspects of the nervous system. Besides regulating human behavior, the nervous system, along with the endocrine system, functions to integrate all of the other systems of the body, so that all systems work together to produce a smoothly functioning organism. This integration is an extremely important part of the overall homeostasis of the body. The nervous system provides the ability for the body to react and adjust to changes both externally and internally in such a manner that normal homeostasis is not disturbed.

This chapter introduces you to the fundamental tissues that compose the organs of the nervous system. It begins with a basic consideration of the principal tissues of both the central and peripheral divisions of the nervous system. From there you will proceed to a consideration of the structure and functioning of neurons, the functional units of the nervous system. Finally, you will be introduced to information processing at the cellular level, or how the nervous system actually functions at its most basic level.

The nervous system is the most complex of all of the organ systems and is usually, for the average person, the most difficult to understand. The secret to mastery of this system is to make sure that you fully comprehend the basic principles that are developed in this chapter. Careful study now will pay big dividends for you in the later, more complex chapters.

CHAPTER OUTLINE

A. Introduction
B. The histology of neural tissue
   1. Neuroglia
      a. Neuroglia of the central nervous system
         (1) Astrocytes
         (2) Oligodendrocytes
         (3) Microglia
         (4) Ependymal cells
      b. Neuroglia of the peripheral nervous system
   2. Neuronal structure
      a. Multipolar neurons
      b. The synapse
      c. Synthesis and transport
C. The physiology of neural tissue
   1. The transmembrane potential
   2. Action potentials and nerve cell membranes
      a. Conduction velocity
   3. The presynaptic membrane and neurotransmitter release
   4. The effects of acetylcholine on the postsynaptic membrane
      a. Temporal summation
      b. Spatial summation
   5. The effects of other important neurotransmitters
   6. Neurons and metabolic processes
D. Information processing
   1. Neurotransmitters and processing at the cellular level
   2. Functional patterns of neural organization
      a. An introduction to reflexes
      b. An introduction to neural processing
E. Clinical patterns
   1. Variations in pH
   2. Variations in temperature
   3. Exposure to chemicals
      a. Drugs that affect membrane properties
      b. Drugs that affect synaptic function
      c. Drugs that alter the rate of transmitter release
      d. Drugs that prevent transmitter activation
      e. Drugs that prevent transmitter binding to receptors
   4. The effects of injury
   5. The effects of diseases

LEARNING ACTIVITIES

Complete the following items by supplying the appropriate word or phrase.

1. The functions of the nervous system include the:
   a. _____ of internal and external environments.
   b. _____ of sensory inputs.

    c. _____ of motor outputs.

    d. _____ of peripheral systems.

2. The spinal cord and brain compose the _____ nervous system.

3. All of the nervous tissue which lies outside of the brain and spinal cord constitutes the _____ nervous system.

4. The _____ is responsible for information transfer in the nervous system.

5. Complete figure 12.1, a typical nerve cell, by supplying the names of each of the four regions. Color each region a different color. Refer to figure 12.1 in your textbook for assistance.

6. The supporting cells of the nervous system are the _____ .

7. The glial cells account for approximately _____ of the volume of the nervous system.

8. Match the neuroglial cells listed below with the appropriate descriptive statement which follows.

a. astrocytes b. oligodendrocytes   c. microglia d. ependymal cells   e. amphicytes   f. Schwann cells

_____ form myelin sheaths for neurons in the PNS

_____ form myelin sheaths for neurons in the CNS

_____ phagocytic cells of the CNS

_____ line the cavities of the CNS

_____ cells that surround the nerve cell bodies in the PNS

_____ responsible for maintaining the blood-brain barrier

_____ form a complete neurilemma in the PNS

_____ provide a three dimensional framework for the CNS

_____ monitor the composition of the CSF

_____ surround the capillaries of the CNS

_____ form scar tissue in the CNS

9. A neuron with a single axon and several dendrites would be a _____ neuron.

10. A _____ neuron has the cell body between a single dendrite and a single axon.

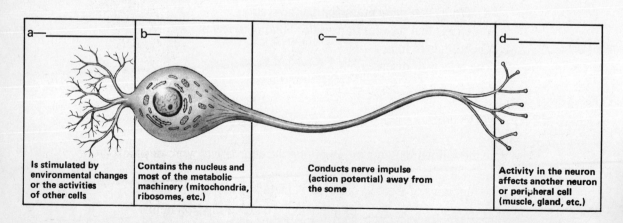

a— _____ b— _____ c— _____ d— _____

Is stimulated by environmental changes or the activities of other cells

Contains the nucleus and most of the metabolic machinery (mitochondria, ribosomes, etc.)

Conducts nerve impulse (action potential) away from the soma

Activity in the neuron affects another neuron or peripheral cell (muscle, gland, etc.)

Figure 12.1

11. A _____ neuron has the axonal and dendritic processes fused.

12. In _____ it is impossible to distinguish axons from dendrites.

13. Skeletal muscles are all innervated by the _____ type neuron.

14. The _____ neurons of the PNS are unipolar.

15. _____ neurons are rare, but have important functions in the eye and the ear.

16. In a typical multipolar neuron, the dendritic complex may account for _____ percent of the surface area of the neuron.

17. Gray area regions of the nervous system consist largely of nerve cell _____.

18. In neurons, ribosomes exist as aggregations that are termed _____.

19. The neurofibrils of a typical neuron are composed of _____.

20. The initial segment of the axon is attached to the axon _____.

21. Branches off of the axon are known as _____.

22. The axon and its branches end in _____ that form expanded synaptic knobs.

23. Label figure 12.2, a typical multipolar neuron, using the terms listed below. Color the various parts. Refer to figure 12.7 in your textbook for assistance.

dendrites	axon hillock	collateral branch	myelin sheath
initial segment	axon	telodendria	synaptic knobs

24. The region where two or more neurons communicate is termed a _____.

25. A _____ junction is a synapse where a neuron communicates with

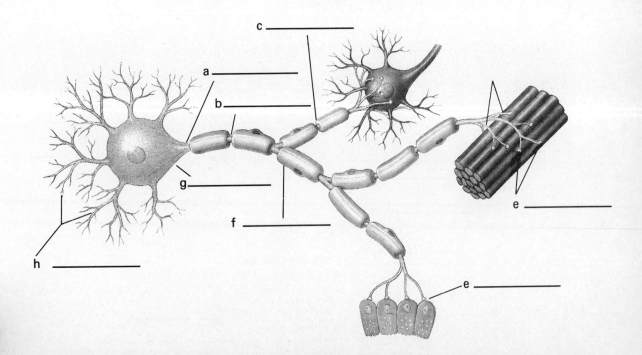

Figure 12.2

another type of cell.

26. Communication across the synapses is one-way, from the presynaptic neuron to the _____ neuron.

27. Communication in a chemical synapse involves the release of a chemical known as a _____ by the presynaptic neuron.

28. The stimulus for relase of a neurotransmitter is the arrival of an _____ .

29. The neurotransmitter is stored in synaptic _____ located in the synaptic knobs.

30. Because the neurotransmitter must diffuse across the synaptic cleft, there is always a synaptic _____ between the arrival of the stimulus at the synaptic knob and the effect on the postsynaptic membrane.

31. Communication across the synapse is one-way because only the _____ neuron releases neurotransmitter, and only the _____ membrane reacts to the presence of neurotransmitter.

32. Label figure 12.3, the synapse, using the terms listed below. Color each of the structures. Refer to figure 12.8 in your textbook for assistance.

neurofilaments	mitochondria	endoplasmic reticulum
synaptic vesicles	synaptic knob	synaptic cleft
postsynaptic membrane	presynaptic membrane	

33. The transmembrane potential in a typical neuron averages about _____ mV.

34. An action potential begins when the transmembrane potential has depolarized to _____ .

35. During the _____ refractory period an action potential may be produced by an abnormally strong stimulus.

36. In neurons, action potentials occur only on the _____ .

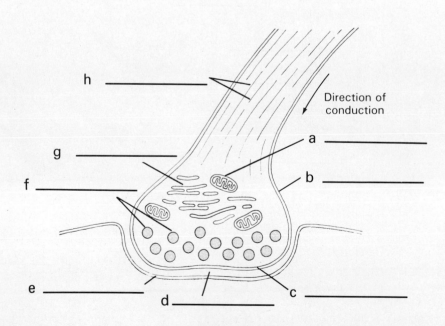

Figure 12.3

37. Action potentials first appear on the neuron when the axon hillock has depolarized to a threshold of about _____ mV.

38. The inability of the dendrites or cell bodies to conduct action potentials means that they always remain _____ to depolarizing or hyperpolarizing stimuli.

39. A neuron which has the axon hillock depolarized to a point which is very close to threshold is said to be _____ because virtually any stimulus will trigger an action potential.

40. Complete figure 12.4, the nerve impulse, by placing the number from the diagram next to the appropriate term listed below. Refer to figure 12.9 in your textbook for assistance.

_____ a. sodium channels open   _____ f. sodium channels close

_____ b. potassium channels open   _____ g. depolarization

_____ c. repolarization   _____ h. threshold

_____ d. absolute refractory   _____ i. relative refractory

_____ e. resting potential   _____ j. gradual depolarization

41. An increase in the diameter of a neuron _____ the rate of impulse conduction.

42. In myelinated fibers, only the _____ depolarize.

43. The jumping from node to node of the action potential in a myelinated fiber is known as _____ conduction.

44. Myelinated fibers with diameters between 4 and 20 micrometers are known as type _____ fibers.

45. Conduction velocity of type B fibers is _____ m per second.

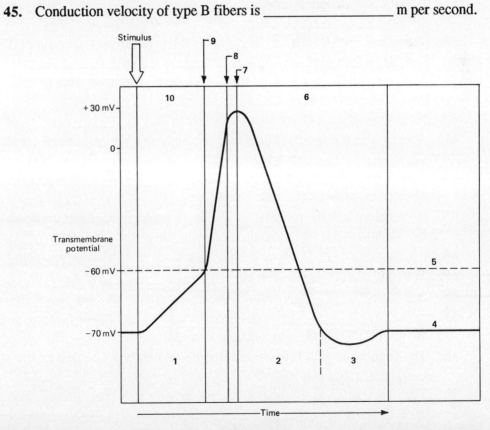

Figure 12.4

46. Most of the motor commands to smooth muscle and glands is conducted by class _____ fibers.

47. _____ synapses use acetylcholine as the neurotransmitter.

48. The actual trigger for the release of acetylcholine is an influx of _____ ions into the cytoplasm following the action potential.

49. The depletion of acetylcholine (or any neurotransmitter) due to repeated rapid firings of the presynaptic neuron will result in synaptic _____.

50. The binding of acetylcholine to receptors on the postsynaptic membrane causes an increase in permeability to _____.

51. The graded depolarization produced on the postsynaptic membrane by the arrival of neurotransmitter is known as an _____.

52. Even if an individual EPSP is too small to initiate an action potential, individual EPSPs can be added up until threshold is reached, a process termed _____.

53. In _____ summation, a single presynaptic neuron fires so rapidly that acetylcholine accumulates to high enought levels to bring the postsynaptic membrane to threshold.

54. When a postsynaptic neuron receives neurotransmitter from several different presynaptic neurons simultaneously, the quantities add up to cause depolarization to threshold, a phenomenon known as _____ summation.

55. Inhibitory neurotransmitters hyperpolarize the postsynaptic membrane thereby creating an _____.

56. Like EPSPs, IPSPs can _____ both temporally and spatially.

57. Besides acetylcholine, another neurotransmitter which is usually excitatory is _____.

58. Three neurotransmitters that are usually inhibitory (produce IPSPs) are _____, _____, and _____.

59. A chemical substance which influences a postsynaptic membrane's response to a neurotransmitter is a _____.

60. Neurons produce ATP exclusively by means of _____ glycolysis.

61. A neuron in a processing center in the brain may be influenced by as many as _____ synapses.

62. An IPSP can cancel out the effects of an _____ if it occurs at the same place on the postsynaptic membrane.

63. A synapse where the presynaptic and postsynaptic membranes are bound together by gap junctions is an _____ synapse.

64. _____ neurons bring internal or external information to the CNS.

65. The body of a sensory neuron lies _____ of the CNS.

66. A _____ neuron controls the activity of a peripheral effector.

67. _____ neurons connect sensory and motor neurons together and are responsible for analysis and coordination.

68. The complexity of a behavioral response is largely a function of the number of _____ involved.

69. The swiftest and most automatic neural responses are termed _____.

70. A _____ reflex has a sensory neuron synapsing directly with a motor neuron.

71. _____ reflexes have one or more interneurons between the sensory and motor neurons.

72. _____ reflexes involve skeletal muscles whereas _____ reflexes are concerned with controlling smooth muscle organs.

73. Interneurons are organized into groups known as neuronal _____.

74. When a single neuron in a neuronal pool controls several other neurons within or without the pool the process is termed _____.

75. When several different neurons control the output of a single neuron the process is known as _____.

76. A _____ circuit will continue to fire indefinitely once it has been stimulated.

77. Match the term listed below with the appropriate descriptive statement that follows.

a. alkalosis   b. acidosis   c. tetrodotoxin   d. saxitoxin   e. caffeine   f. botulinus toxin   g. anticholinesterase   h. atropine   i. nicotine

_____ prevents acetylcholine from binding to the postsynaptic membrane

_____ increases the excitability of the nervous system

_____ blocks the release of ACh by the presynaptic neuron

_____ fish toxin which blocks sodium channels

_____ a class of compounds that prevent transmitter inactivation

_____ substance that causes paralytic shellfish poisoning

_____ substance binds to receptor sites thereby stimulating the postsynaptic membrane

_____ makes neurons more sensitive to depolarizing stimuli

_____ this condition can result in coma

SELF TEST

Circle the correct answer to each question.

1. The neuroglial cell which plays a major role in establishing the blood-brain barrier is the
   a. astrocyte.   b. oligodendrocyte   c. ependymal cell.   d. microglia   e. Schwann cell.

2. The neuroglial cell which forms myelin in the CNS is the
   a. astrocyte.   b. oligodendrocyte.   c. ependymal cell.   d. microglia.   e. Schwann cell.

3. The cell that lines the hollow cavities of the CNS is the
   a. amphicyte.   b. ependymal cell.   c. Schwann cell.   d. microglia.
   e. oligodendrocyte.

4. The cell which is responsible for forming the neurilemma in the PNS is the
   a. amphicyte.   b. ependymal cell.   c. Schwann cell.   d. astrocyte.   e. microglia.

5. A neuron with one axon and seven dendrites would be classified as
   a. unipolar.   b. bipolar.   c. anaxonic.   d. multipolar.   e. afferent.

6. Aggregations of ribosomes that occur in the neuron cell body are termed
   a. perikaryons.   b. lipofuscins.   c. Nissl bodies.   d. telodendria.   e. synaptic knobs.

7. The neurotransmitter of a neuron is contained in the
    a. dendrites.   b. initial segment.   c. axon hillock.   d. synaptic vesicles.   e. soma.

8. In a neuron, an action potential occurs on the
    a. dendrites.   b. cell body.   c. axon.   d. nucleus.   e. more than one of the above
    is correct.

9. Conduction from node of Ranvier to node of Ranvier is known as
    a. continuous.   b. slow.   c. saltatory.   d. atypical.   e. none of the above.

10. The fastest conducting fibers are the
    a. class A.   b. class b.   c. class c.   d. class d.   e. class E.

11. Fibers which release acetylcholine are termed
    a. adrenergic.   b. cholinergic.   c. electrical.   d. inhibitory.   e. more than one of
    the above is correct.

12. The trigger for the release of neurotransmitter is the influx of _____
    into the neuron's cytoplasm.
    a. sodium.   b. potassium.   c. magnesium.   d. calcium.   e. chloride.

13. The normal resting potential of a neuron is -70 mV. If an IPSP were placed on this
    membrane it would most closely approach _____ mV.
    a. -60   b. -65   c. +25   d. 0   e. -80

14. An example of an inhibitory neurotransmitter is
    a. acetylcholine.   b. norepinephrine.   c. noradrenalin.   d. dopamine.   e. none of
    the above.

15. The most numerous type of functional neuron found in the body is the
    a. afferent.   b. motor.   c. sensory.   d. efferent.   e. interconnecting.

CHAPTER 13

The Nervous System:
The Spinal Cord and
Spinal Nerves

OVERVIEW

This chapter continues the survey of the nervous system by the spinal cord and spinal nerves. The spinal cord is part of the CNS in an anatomical sense but it is fully integrated with the PNS in a functional sense. Therefore the details of the nerves that exit the cord will also be discussed in this unit. The discussion begins with a survey of the anatomy of the cord and the distribution of the various kinds of neurons within it. This includes not only the cord proper, but the protective covering of the cord, the meninges, a series of membranes which surround it. Once the organization of the cord has been described you will then be introduced to the peripheral nervous system. The first part of that introduction will involve the spinal nerve organization. From there you will proceed to the study of the nerve plexuses, complex interactions of various spinal nerves that ultimately give rise to many of the peripheral nerves of the body. This will complete the anatomic survey for this chapter and you will now proceed to an examination of the integrated functioning of the cord and PNS, specifically, the spinal reflexes. Finally you will examine how the basic spinal reflexes are integrated into higher centers of control.

This material may appear complex to you, but if you will keep in mind the basic principles of neural tissue which you learned in the last chapter you will find that the current material is much easier. In the previous chapter you learned the principles upon which neural circuits operate. In this chapter you are examining the intact circuits which all operate by the basic principles that govern all neural tissue.

CHAPTER OUTLINE

A. Introduction
B. Anatomical organization of the spinal cord and peripheral nervous system

1. The distribution of neurons and axons
2. Gross anatomy of the spinal cord
   a. The spinal meninges
3. Sectional anatomy of the spinal cord
   a. Gray matter organization
   b. White matter organization
4. The peripheral nervous system
   a. Spinal nerves
   b. The peripheral distribution of spinal nerves
   c. Nerve plexuses
C. Functional integration of the spinal cord and peripheral nervous system
   1. Spinal reflexes
      a. Monosynaptic reflexes
      b. Polysynaptic reflexes
         (1) The tendon reflex
         (2) The withdrawal reflexes
         (3) The crossed extensor reflexes
   2. Integration with higher centers
D. Clinical patterns
   1. Impairment of peripheral nerve function
   2. Spinal meningitis
   3. Shingles and palsies
   4. Impairment of spinal cord function

LEARNING ACTIVITIES

Complete the following items by supplying the appropriate word or phrase.

1. In addition to passing information to the brain, the spinal cord also serves as an _____ center.

2. Within the CNS, a _____ is a group of neurons with distinct anatomical boundaries.

3. Gray matter centers within the spinal cord are known as _____, while similar masses of cell bodies in the PNS are termed _____.

4. _____ provide information about the outside world.

5. _____ provide information as to the position of the skeletal muscles and joints.

6. _____ monitor internal operations.

7. Match the terms below with the appropriate descriptive statements which follow.

   a. sensory neuron b. motor neuron c. interneuron d. receptor e. somatic motor neuron f. visceral motor neuron g. tract h. column i. pathway

   _____ A sensory structure that detects changes in the environment.

   _____ The body of this motor neuron lies inside of the CNS.

   _____ These neurons are found completely inside of the CNS and are responsible for integration.

   _____ Bundles of axons in the CNS that share common origins, destinations, and functions.

   _____ These motor neurons have a neuron that arises from a ganglion.

_____ A neuron which stimulates or modifies the activity of a peripheral tissue.

_____ These are bundles of tracts.

_____ These consist of nuclei and tracts that link the brain to the rest of the body.

_____ The neuron which connects a receptor with the spinal cord or brain.

8. The _____ is a shallow longitudinal groove on the posterior surface of the spinal cord.

9. The deep crease on the anterior side of the spinal cord is the _____.

10. Increased amounts of gray matter create _____ in the segments of the cord concerned with control of the limbs.

11. The spinal cord can be divided into _____ segments.

12. Each segment gives rise to a pair of _____ nerves.

13. The dorsal root ganglion contains the cell bodies of the _____ neurons.

14. The ventral root of each spinal nerve contains the axons of _____ neurons.

15. The conical shaped tip of the spinal cord is termed the _____.

16. The _____ is a filamentous extension of the cord that continues to the second sacral vertebra.

17. The outermost of the meninges is the _____.

18. The space between the dura mater and the walls of the vertebral canal is the _____ space.

19. The subdural space separates the dura mater from the _____.

20. The subarachnoid space is filled with _____ which serves as a shock absorber for the CNS.

21. The innermost meningeal membrane is the _____.

22. The pia-arachnoid is connected to the dura mater by the _____ ligaments.

23. The spinal cord ends at about the _____ lumbar vertebra.

24. Label figure 13.1, the spinal cord, with the terms listed below. Color the various structures. Refer to figures 13.2 and 13.3 in your textbook for assistance.

anterior median fissure	posterior median sulcus	dorsal root
ventral root rootlets	meninges pia mater	arachnoid
mater spinal nerve	gray matter white matter	dura mater

25. The projections of gray matter in the cord which project towards the surface are the _____.

26. The cell bodies of the spinal cord are organized into _____.

27. The _____ gray horns contain somatic and visceral sensory nuclei.

28. The _____ gray horns contain somatic motor nuclei.

29. The _____ gray horns contain visceral motor nuclei.

30. Axons which cross over from one side of the cord to the other do so through the gray _____.

31. The white matter of the cord is divided into a half-dozen _____.

32. The columns contain _____.

33. _____ tracts convey sensory information while _____ tracts convey motor signals.

Label: anterior median fissure   posterior median sulcus   dorsal root   ventral root   rootlets   meninges
pia mater   arachnoid mater   spinal nerve   gray matter   white matter   dura mater

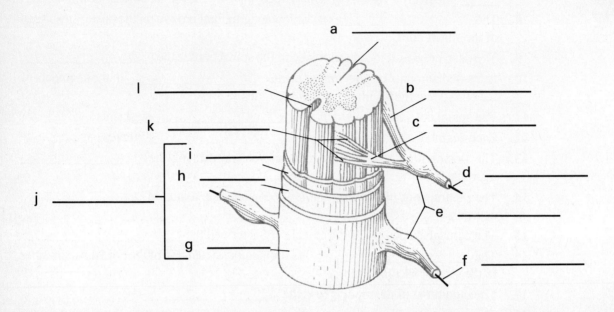

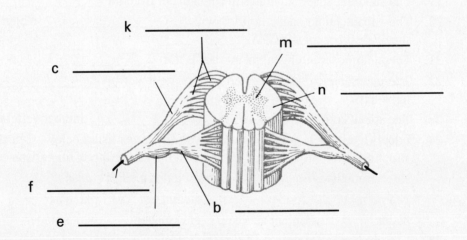

Figure 13.1

34. Label figure 13.2 , sectional organization of the cord, using the terms listed below.
    Color each section. Refer to figure 13.5 in your textbook for assistance.

posterior gray commissure	posterior median sulcus	dorsal root
somatic	visceral	sensory
anterior gray commissure	anterior median fissure	anterior gray horn
lateral gray horn	posterior gray horn	posterior white column
lateral white column	anterior white column	motor
anterior white commissure	ventral root	

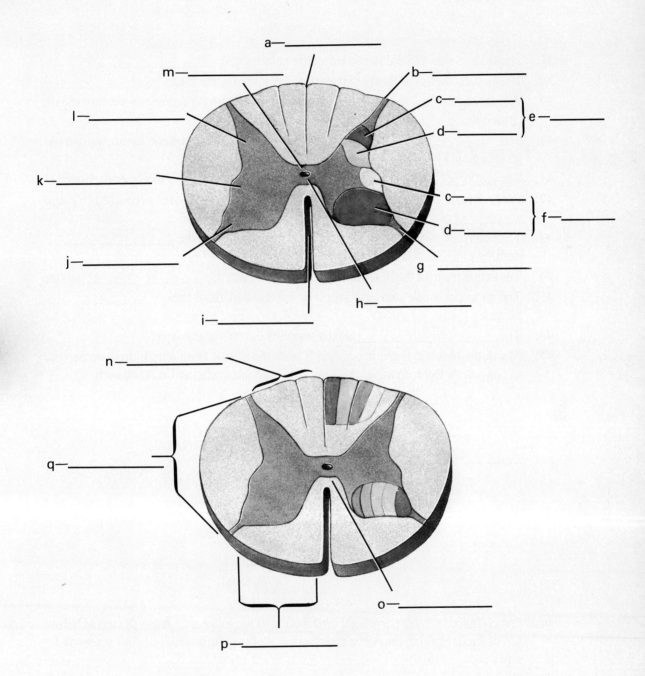

Figure 13.2

35. _____ is the introduction of radiopaque dyes into the CSF.

36. There are 31 pairs of spinal nerves. List the number of pairs per region.

    _____ cervical

    _____ thoracic

    _____ lumbar

    _____ sacral

    _____ coccygeal

37. The fifth thoracic spinal nerve would be abbreviated _____.

38. The connective tissue sheath that surrounds a spinal nerve is the _____.

39. Within a nerve, each individual fiber is surrounded by a connective tissue sheath termed the _____.

40. The _____ ramus carries preganglionic fibers to an autonomic ganglion.

41. The _____ ramus carries postganglionic fibers to visceral effectors.

42. Innervation to the skin and muscles of the back are provided by the _____ ramus.

43. The ventrolateral body wall and the limbs are supplied by the _____ rami.

44. The ventral rami of the first four cervical nerves form the _____ plexus.

45. The shoulder girdle and arm receives innervation from the _____ plexus.

46. The _____ plexus supplies the pelvis and legs.

47. Match the nerve(s) listed in column B with the plexus from which they are derived in column A. Refer to tables 13.1 and 13.2 in your textbook for assistance.

A	B
_____ cervical plexus	a. radial
	b. femoral
_____ brachial plexus	c. phrenic
	d. obturator
_____ lumbar plexus	e. iliohypogastric
	f. median
_____ sacral plexus	g. gluteal
	h. ulnar
	i. pudenal
	j. sciatic

48. Label figure 13.3, the peripheral distribution of a spinal nerve, using the terms below. Color each region of the nerve distribution. Refer to figure 13.7 in your textbook for assistance.

    dorsal ramus   ventral ramus   spinal nerve   dorsal root
    ganglion   white ramus   gray ramus   autonomic ganglion
    autonomic nerve

49. One of the few examples of a monosynaptic reflex is the _____ reflex.

50. In the withdrawal reflex, motor neurons to the flexor muscle are excited while those to the extensor muscle are _____.

51. The excitation and inhibition of motor neurons that occurs during the withdrawal reflex is brought about by _____ neurons.

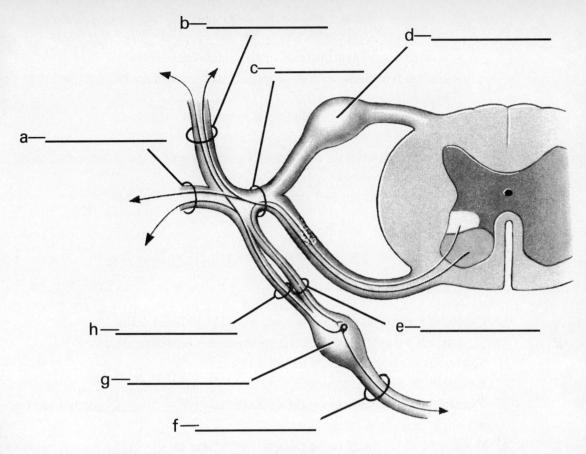

Figure 13.3

52. A person has just stepped on a sharp object with a bare right foot. The person immediately flexes the right leg to remove the foot from the source of pain, and simultaneously extends the left leg for balance. In the following statements about this situation, supply the missing information concerning the neural pathways involved.

   a. The reflex was initiated by a pain _____.

   b. The impulse from the receptor was conveyed back to the right side of the cord by a _____ neuron which then synapsed with four interconnecting neurons.

   c. The interneuron which acted upon the motor neurons which innervated the right leg flexors was _____ (excitatory or inhibitory).

   d. The interneuron which acted upon the motor neurons which innervated the left leg flexors was _____.

   e. The interneuron which acted upon the motor neurons which innervated the right leg extensors was _____.

   f. The interneuron which acted upon the motor neurons which innervated the left leg extensors was _____

   g. The motor neurons which terminated in the muscles of both legs were _____.

53. Although spinal reflexes are not under voluntary control, the movements which they cause can be activated conciously via the higher _____ found in the brain.

54. Match the clinical condition listed below with the appropriate descriptive statement which follows.

a. lead poisoning   b. spinal meningitis   c. sciatica   d. peroneal palsy   e. multiple sclerosis   f. spina bifida

_____ is often caused by a distorted lumbar intervertebral disk

_____ can result in destruction of peripheral axons resulting in sensory and motor impairment

_____ an inflammation of the meningeal membranes

_____ failure of the vertebral laminae to unite

_____ sensory loss from the top of the foot and side of the lower leg due to sitting with the legs crossed for extended periods

_____ muscular paralysis and sensory loss due to demyelination

SELF TEST

Circle the correct answer to each question.

1. Groups of nerve cells with distinct anatomical boundaries define
   a. tracts.   b. nerves   c. centers.   d. epineurium.   e. endoneurium.

2. The bodies of sensory neurons lie in
   a. nuclei.   b. centers.   c. ganglia.   d. tracts.   e. nerves.

3. Visceral motor pathways always involve at least _____ motor neurons.
   a. 1   b. 2   c. 3   d. 4   e. 5

4. Bundles of axons found in the CNS are equivalent to _____ found in the PNS.
   a. tracts.   b. nerves.   c. ganglia.   d. Schwann cells.   e. amphicytes

5. There are _____ segments in the spinal cord.
   a. 12   b. 5   c. 8   d. 31   e. 42

6. The dorsal root of a spinal nerve contains
   a. centers.   b. nucleic   c. tracts.   d. sensory fibers.   e. motor fibers.

7. The outermost of the meninges is the
   a. dura mater.   b. pia mater.   c. subdural space.   d. arachnoid.   e. subarachnoid space.

8. The anterior gray horns contain
   a. sensory neurons.   b. motor neurons.   c. interneurons.   d. ganglia.   e. Schwann cells.

9. The connective tissue layer that compartmentalizes axons into bundles within a nerve is the
   a. epineurium.   b. perineurium.   c. endoneurium.   d. neurofibril.   e. none of the above.

10. The median nerve arises from the
    a. cervical plexus.   b. brachial plexus.   c. lumbar plexus.   d. sacral plexus.   e. dorsal ramus.

11. In a spinal reflex involving reciprocal innervation, if an extensor is inhibited, then its antagonistic flexor will be
    a. excited.   b. inhibited.   c. either inhibited or excited depending upon the nature of the stimulus.   d. not effected.   e. none of the above.

12. In spinal reflexes, all inhibition or excitation of motor neurons is carried out by the

   a. sensory neurons.   b. interconnecting neurons.   c. motor neurons.   d. afferent neurons.   e. efferent neurons.

13. Higher centers may influence reflex movement by

   a. stimulating the sensory neurons.   b. stimulating the motor neurons.   c. stimulating the interneurons.   d. altering ganglia activity.   e. none of the above.

14. Which of the following reflexes employs reciprocal innervation?

   a. load reflex   b. postural reflexes   c. patellar reflex   d. stretch reflex
   e. withdrawal reflex

15. Demyelination of tracts results in

   a. spina bifida.   b. multiple sclerosis.   c. shingles.   d. meningitis.   e. sciatica.

CHAPTER 14

The Nervous System: The Brain and Cranial Nerves

OVERVIEW

In chapter 13 you began a study of both the central and peripheral nervous systems by examining the spinal cord and the nerves which it gives rise to. In this chapter this study is continued by examining the remaining parts of both divisions, the brain of the CNS, and the cranial nerves of the PNS. The chapter begins with an examination of the cranial meninges and how they differ from their spinal equivalents. This is followed by an examination of the origin, circulation, and function of the cerebrospinal fluid. Next is an examination of the gross anatomy of the brain. The brain is organized into five regions (telencephalon, diencephalon, mesencephalon, metencephalon, and myelencephalon). The major structures and the functions of each of the regions will be surveyed. Finally the twelve pairs of cranial nerves will be examined. The cranial nerves are those parts of the PNS which originate directly from the brain.

Although the spinal cord and nerves were considered in a separate chapter, it is important to remember that they are continuous with the brain and with it form an integrated functional unit.

CHAPTER OUTLINE

A. Introduction
B. The cranial meninges
C. Cerebral fluid formation and circulation
D. Gross anatomy and organization of the brain

1. The cerebrum (telencephalon)
   a. The cerebral cortex
   b. The central white matter
   c. The cerebral nuclei
   d. The limbic system
2. The diencephalon
   a. The thalamus
   b. The hypothalamus
3. The mesencephalon or midbrain
4. The metencephalon
   a. The cerebellum
   b. The pons
5. The medulla (myelencephalon)
E. The cranial nerves
   1. Olfactory nerve
   2. Optic nerve
   3. Oculomotor nerve
   4. Trochlear nerve
   5. Trigeminal nerve
   6. Abducens nerve
   7. Facial nerve
   8. Acoustic nerve
   9. Glossopharyngeal nerve
   10. Vagus nerve
   11. Spinal accessory nerve
   12. Hypoglossal nerve
   13. A summary of cranial nerve branches and functions
F. Clinical patterns
   1. Headaches
   2. Brain dysfunction
      a. Cranial trauma
      b. Impairment of circulation: cerebrovascular disease
   3. CNS infections and inflammations

LEARNING ACTIVITIES

Complete the following items by supplying the appropriate word or phrase.

1. In the brain, when gray matter is layered over white matter, a neural _____ is formed.
2. The brain can communicate with peripheral structures via tracts that pass down the cord or directly by means of the _____ nerves.
3. The five regions of the adult brain, from anterior to posterior, are the _____, _____, _____, _____, and _____.
4. The blood vessels known as the _____ lie between the two layers of the dura mater that surround the brain.
5. The extension of the dura mater that projects into the longitudinal fissure between the cerebral hemispheres is the _____.

6. The _____ is the dura mater extension that separates the cerebellum from the cerebrum.

7. The _____ are networks of capillaries found in the ventricles of the brain which produce CSF.

8. In addition to protection, CSF acts as an important _____ medium for metabolites and other substances.

9. Ventricles I and II are found in the _____.

10. The third ventricle is found in the _____.

11. The third ventricle is connected to the fourth via the _____.

12. CSF formed in the ventricles enters into the subarachnoid space by means of three holes in the roof of the _____ ventricle.

13. CSF is returned to the circulation by the _____ villi.

14. Label figure 14.1, the brain and ventricles, using the terms listed below. Color the various structures. Refer to figures 14.1 and 14.3 in your textbook for assistance.

fourth ventricle	cerebellum	cerebral aqueduct
lateral ventricle	cerebral hemisphere	interventricular foramen
third ventricle	pons	medulla

15. In figure 14.2, circulation of cerebrospinal fluid, color the pathway that CSF follows. Begin with the lateral ventricle and follow the arrows. Refer to figure 14.3 in your textbook for assistance.

16. The _____ are folds in the cerebral cortex.

17. The gyri of the cortex are separated from one another by grooves termed _____.

18. The two cerebral hemispheres are separated from one another by the _____ fissure.

19. List the five lobes that each cerebral hemisphere is divided into.

    _____

    _____

    _____

    _____

    _____

20. The frontal lobes are separated from the parietal lobes by the _____.

21. The primary motor area is located on the _____ gyrus.

22. The primary sensory area is located on the _____ gyrus.

23. Sensory and motor areas of the cortex are connected by _____ areas.

24. Portions of the same hemisphere are connected together by _____ fibers.

25. _____ fibers connect the two hemispheres together.

26. The cerebral cortex is linked to the remainder of the brain and nervous system by _____ fibers.

27. Two prominent bundles of commissural fibers are the _____ and the _____.

28. Most of the projection fibers are contained in a structure known as the _____ capsule.

29. Match the terms listed below with the appropriate descriptive statement which follows.

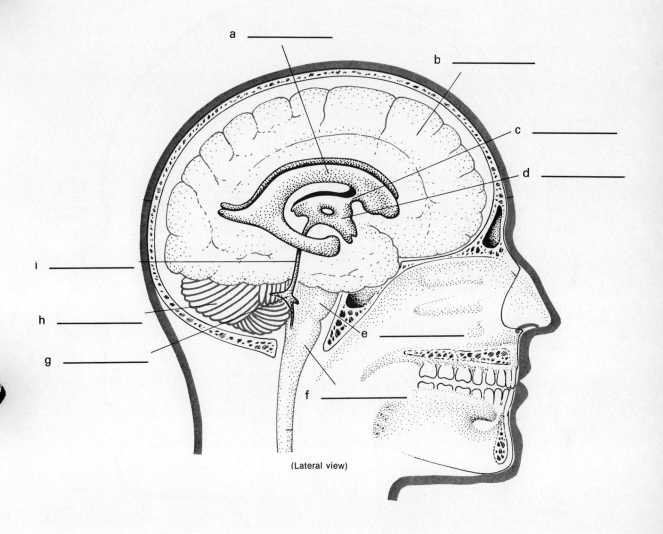

(Lateral view)

Figure 14.1

a. basal nuclei (ganglia)   b. caudate nucleus   c. amygdaloid body   d. lentiform nucleus   e. limbic system   f. hippocampus   g. limbic lobe   h. anterior nucleusim i. corpus striatum

_____ functions in the processing of memories, creation of emotional states, drives, and associated behavior

_____ this structure includes the putamen and globus pallidus

_____ caudate and lentiform nucleus are components of this

_____ play a role in adjusting motor activities

_____ appears to be important in the storage of long term memory

_____ nucleus with a massive head and a curving tail

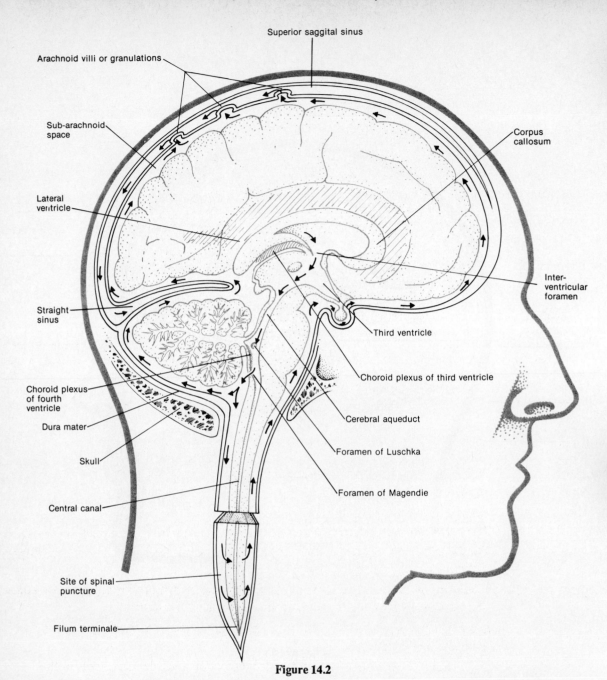

Superior saggital sinus

Arachnoid villi or granulations

Sub-arachnoid
space

Lateral
ventricle

Straight
sinus

Choroid plexus
of fourth
ventricle

Dura mater

Skull

Central canal

Site of spinal
puncture

Filum terminale

Corpus
callosum

Inter-
ventricular
foramen

Third ventricle

Choroid plexus of third ventricle

Cerebral aqueduct

Foramen of Luschka

Foramen of Magendie

Figure 14.2

_____ acts as an interface between the limbic system and the cerebrum

_____ gyri that curve along the corpus callosum on to the surface of the temporal lobe

_____ structure that relays visceral sensations from the hypothalamus to the cerebrum

30. Label figure 14.3, the limbic system, using the terms listed below. Color each structure. Refer to figure 14.7 in your textbook for assistance.

fornix   corpus callosum   central sulcus   cingulate gyrus
temporal lobe   hippocampus p   arahippocampal gyrus
mamillary body

31. The three major structures or regions of the diencephalon are the _____, _____, and _____.

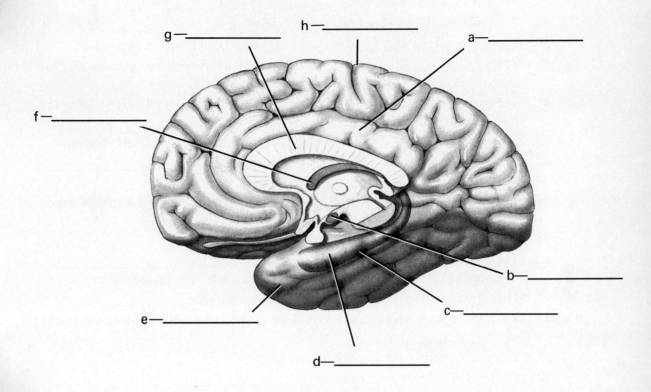

Figure 14.3

32. The pineal body or gland is an _____ gland.

33. The principal role of the thalamus is the processing of _____ information.

34. Match the thalamic nuclei, listed below with the appropriate descriptive statement that follows. Refer to table 14.4 in your textbook for assistance.

    a. anterior nuclei   b. ventral nuclei   c. pulvinar nucleus   d. lateral geniculates   e. medial geniculates

    _____ integrates sensory information for projection to association areas of the cerebral cortex

    _____ parts of the limbic system

    _____ relay information from the cerebellum and cerebral nuclei to motor areas of the cerebral cortex

    _____ projects auditory information to the auditory cortex

    _____ projects visual information to the visual cortex

35. Write in the part of the hypothalamus that is responsible for each of the functions listed below. Refer to table 14.5 in your textbook for assistance.

    _____ control feeding reflexes

    _____ zone that is responsible for controlling body temperature

    _____ nucleus that secretes antidiuretic hormone

    _____ control heart rate and blood pressure

    _____ nucleus that secretes the hormone oxytocin

36. The hypothalamus plays a major role in internal integration in itself and by integrating with the _____ system.

37. The tectum of the mesencephalon is formed by the _____.

38. The superior colliculi play a major role in relaying _____ inputs.

39. The _____ receive auditory data from nuclei in the medulla.

40. The _____ nucleus issues involuntary motor commands after integrating information from the cerebrum and cerebellum.

41. The nucleus known as the _____ nigra plays a role in regulating the motor output of the cerebral nuclei.

42. The _____ are the principal white matter of the mesencephalon.

43. The two major structures found in the metencephalon are the _____ and the _____.

44. The cerebellar hemispheres of the posterior lobe are separated by a strip of cortex known as the _____.

45. The cerebellum directs _____ reflexes.

46. The cerebellum coordinates the conscious and unconscious _____ commands of the cerebral hemispheres, cerebral nuclei, and brain stem.

47. The cerebellar peduncles connect the cerebellum with the _____.

48. The inferior cerebellar peduncle permits the cerebellum to communicate with nuclei in the _____.

49. The pons contains a number of longitudinal _____.

50. The myelencephalon consists of the _____.

51. All communication between the brain and the spinal cord involve the _____.

52. The _____ nucleus of the medulla relays information to the cerebellum.

53. There are motor and sensory nuclei for _____ cranial nerves in the medulla.

54. The _____ formation in the medulla contains autonomic reflex centers.

55. Label figure 14.4, sagittal section of the brain, using the terms listed below. Color the various structures. Refer to figure 14.1 in your textbook for assistance.

spinal cord	medulla oblongata	pons	temporal lobe
hypophysis	hypothalamus	anterior commissure	frontal lobe
corpus callosum	fornix	parietal lobe	thalamus
occipital lobe	pineal gland	cerebellum	

56. Write the name of the cranial nerve in front of the statement that describes its major function (innervation). Refer to table 14.9 of your textbook for assistance.

_____ supplies many of the visceral organs in the thorax and abdominal regions

_____ controls the tongue muscles

_____ receives stimuli from the olfactory epithelium

_____ controls the lateral rectus muscle

_____ controls the superior oblique muscle

_____ controls the inferior, medial, and superior rectus muscles

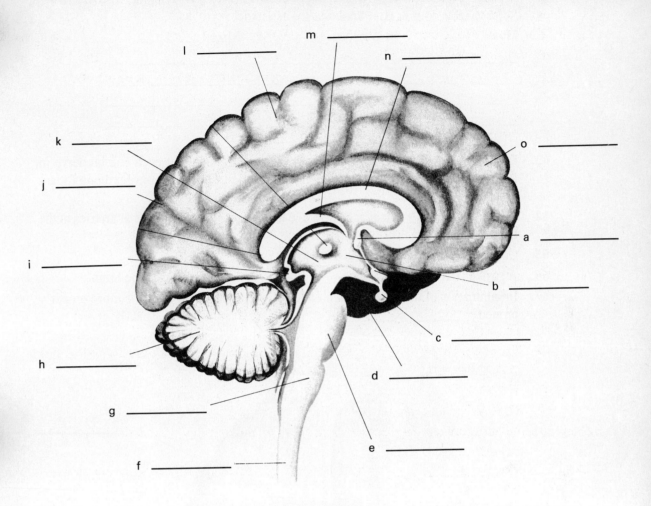

Figure 14.4

_____ among others, it innervates the sternocleidomastoid and
trapezius muscles

_____ innervates the cochlea and vestibule

_____ termed the "toothache nerve" because it receives sensory
input from the gums and teeth

_____ sensory from the posterior region of the tongue, pharynx, and
palate

_____ sends motor fibers to most of the muscles involved in facial
expression

57. The three major branches of the trigeminal nerve are the _____,
_____, and _____.

58. The two major branches of the acoustic nerve are the _____ and the
_____.

59. The sensory ganglion of the trigeminal nerve is the _____.

60. The _____ is the sensory ganglion of the facial nerve.

61. The petrosal ganglion is sensory for the _____ nerve.

62. In the three columns below, list the cranial nerves which fit under each heading. Refer to table 14.9 in your textbook for assistance.

Motor   Sensory   Mixed

_____   _____   _____

_____   _____   _____

_____   _____   _____

_____   _____

_____

63. Label figure 14.5, the cranial nerves, by writing the name of each cranial nerve in the space provided. Color each nerve. Refer to figure 14.13 in your textbook for assistance.

64. _____ headaches are due to changes in the diameter of arteries in the brain.

65. Contusions usually occur with severe _____.

66. A stroke is due to the loss of _____ to a region of the brain.

67. Inflammation of the brain is often called "sleeping sickness" but it is more properly termed _____.

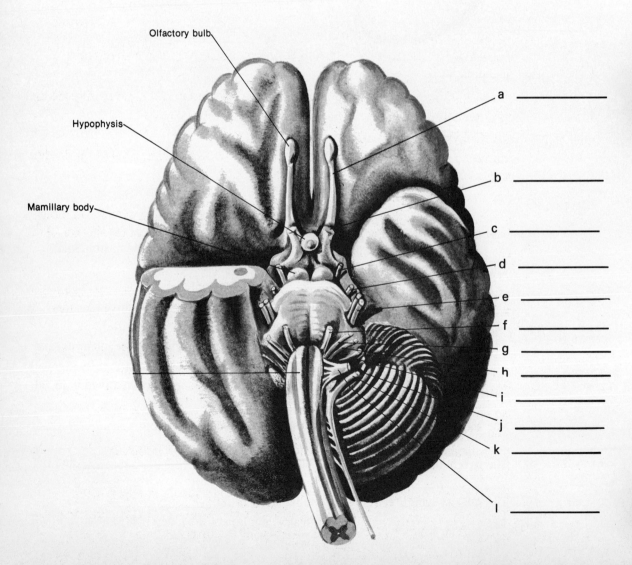

Olfactory bulb

Hypophysis

Mamillary body

a _____

b _____

c _____

d _____

e _____

f _____

g _____

h _____

i _____

j _____

k _____

l _____

Figure 14.5

SELF TEST

Circle the correct answer to each question.

1. The dural partition that occurs between the cerebellum and cerebrum is the
   a. falx cerebri.   b. falx cerebelli.   c. tentorium cerebelli.   d. epidural space.
   e. subdural space.

2. The midbrain region is the
   a. telencephalon.   b. diencephalon.   c. mesencephalon.   d. metencephalon.
   e. myelencephalon.

3. The third ventricle is found in the
   a. cerebral hemispheres.   b. diencephalon.   c. mesencephalon.   d. metencephalon.   e. cerebellum.

4. CSF is produced by the
   a. arachnoid villi.   b. denticulate ligaments.   c. choroid plexuses.   d. pia mater.
   e. dura mater.

5. The central sulcus separates the frontal lobe from the
   a. occipital lobe.   b. parietal lobe.   c. insula.   d. temporal lobe.   e. none of the
   above.

6. Which of the following is not a cerebral nucleus?
   a. caudate   b. putamen   c. globus pallidus   d. claustrum   e. paraventricular

7. The limbic system functions in
   a. gross motor control of the legs.   b. equilibrium and balance.   c. processing visual
   information.   d. emotional states.   e. inhibitory motor activity.

8. The reticular formation is found
   a. in the cerebrum.   b. in the diencephalon.   c. in the mesencephalon   d. in the
   length of the brain stem.   e. in the medulla.

9. The thalamus functions as a
   a. major sensory processing center.   b. major motor processing center.   c. motor
   inhibitory center.   d. motor excitatory center.   e. more than one of the above is correct.

10. The hypothalamus functions in
    a. temperature regulation.   b. thirst sensation.   c. hunger sensation.   d. coordina-
    tion of the autonomic nervous system.   e. more than one of the above is correct.

11. The inferior colliculi function to
    a. relay visual information.   b. relay auditory information.   c. relay motor signals.
    d. accept CSF from the arachnoid space.   e. stimulate the cerebellum.

12. The cerebellum controls _____ reflexes.
    a. cardiovascular   b. respiratory   c. postural   d. endocrine   e. auditory

13. The fifth cranial nerve is the
    a. abducens.   b. oculomotor.   c. vagus.   d. trigeminal.   e. trochlear.

14. The lateral rectus muscle receives innervation from the
    a. abducens.   b. trochlear.   c. oculomotor.   d. trigeminal.   e. hypoglossal.

15. The _____ nerve carries autonomic fibers to the abdominal viscera.
    a. vagus   b. hypoglossal   c. glossopharyngeal   d. spinal accessory   e. facial

CHAPTER 15
The Nervous System: Information Processing and Higher-Order Functions

OVERVIEW

The previous chapters in this unit have introduced you to the basic principles of neural tissue functioning and the anatomy of the nervous system. You have also seen how the simplest neural circuits, the spinal reflexes, operate. This chapter is concerned almost exclusively with the functioning of the brain, the higher-order functions. It begins with a consideration of the neurotransmitters of the brain and then proceeds into the principles of information processing for both sensory and motor pathways. Once these basic principles have been established consideration is then directed toward the higher-order functions such as sleep, arousal, memory and learning, and the integration of information.

Although these higher functions of the brain are very complex, they are still based upon the principles upon which the spinal reflexes utilize. There must be a receptor, sensory pathway, integrator, motor pathway, and finally, an effector. The major difference is that the intergrator portion is now much more complex and involves many more interconnecting neurons than do even the most complex spinal reflexes. If you keep these facts in mind as you progress through this chapter it will make understanding of the material far easier.

CHAPTER OUTLINE

A. Introduction
B. Neurotransmitters in the brain
C. Principles of information processing
    1. Sensory and motor pathways

   a. Sensory pathways
    (1) The posterior column pathway
    (2) The spinothalamic pathway
   b. Motor pathways
    (1) The pyramidal system
    (2) The extrapyramidal system
     (a) The cerebral nuclei
    (3) The role of the cerebellum
  2. Higher order functions
   a. The asleep, awake, alert, and attentive states
    (1) Brain activity and the electroencephalogram
    (2) The unconscious and asleep states
   b. Memory and learning
   c. The integration of information
   d. Hemispheric dominance
   e. Brain asymmetries
   f. Personality and other mysteries
 D. Clinical patterns
  1. Cerebral palsy
  2. Seizures
  3. Genetically programmed disorders
  4. Altered states of awareness

LEARNING ACTIVITIES

Complete the following items by supplying the appropriate word or phrase.

1. Neural processing by both the brain and cord involves the following sequence.
   input - divergence - _____ - convergence - output

2. Activities in the brain differ from those in the cord in that there are many more synapses per neuron and a much wider variety of_____.

3. There are _____ different neurotransmitters in the brain.

4. The neurotransmitters _____ and _____ are produced by neurons in the reticular formation.

5. _____ is manufactured by neurons in the substantia nigra.

6. Two neuromodulators that reduce pain are _____ and _____.

7. Ascending sensory information may be processed by the _____, _____, and _____ before reaching the cerebral cortex.

8. Higher, conscious centers often perform voluntary actions by activating preestablished _____ patterns.

9. _____ is the condition whereby the brain fails to develop higher than the mesencephalon.

10. Anencephalic babies may appear normal at first because so much of early normal behavior is regulated by the _____.

11. Pathways include the peripheral nerves, tracts, and _____ that connect higher centers with the rest of the body.

12. Identify the major tracts in figure 15.1, ascending and descending tracts of the spinal cord, by writing the letter that indicates the tract in front of the appropriate tract name in the list below. Color each tract. Refer to figure 15.1 in your textbook for assistance.

    1. _____ tectospinal tract
    2. _____ rubrospinal tract
    3. _____ lateral spinothalamic tract
    4. _____ anterior corticospinal tract
    5. _____ lateral corticospinal tract
    6. _____ posterior spinocerebellar tract
    7. _____ anterior spinothalamic tract
    8. _____ vestibulospinal tract
    9. _____ anterior spinocerebellar tract
    10. _____ reticulospinal tract
    11. _____ fasciculus gracilis
    12. _____ fasciculus cuneatus

13. A tract whose name begins with _____ begins in the spinal cord and is therefore sensory.

14. A tract whose name ends in _____ terminates in the spinal cord and therefore must be motor.

15. The posterior columns contain the tracts known as the _____ and _____.

16. The fasciculus gracilis synapses with the nucleus gracilis in the _____.

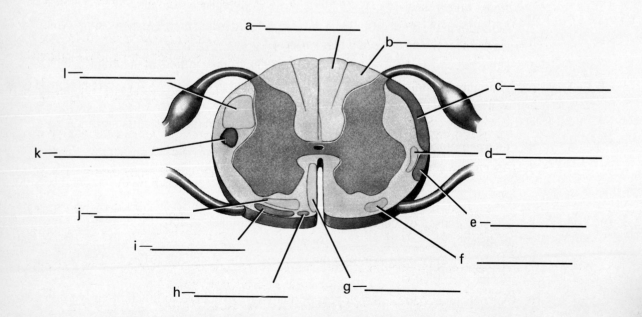

Figure 15.1

17. The _____ transmits sensations of touch, pressure, pain, and temperature.

18. A pain on the left side of the body would be interpreted by the _____ side of the brain.

19. The spinothalamic tracts synapse in the _____ and the postsynaptic fibers then terminated in the primary sensory cortex.

20. Proprioceptive information concerning the position of muscles, tendons, and joints is carried by the _____ tracts.

21. Voluntary motor commands are routed over the _____ system.

22. The _____ tracts terminate on the nuclei of cranial nerves.

23. In the medulla 85% of the corticospinal fibers cross over to form the _____ corticospinal tract.

24. The fibers of the anterior corticospinal tracts cross over within the anterior white _____.

25. If you pick up an object with your right hand, the impulses for this action were initiated by the motor cortex on the _____ side of the brain.

26. List the four tracts of the extrapyramidal system

    _____

    _____

    _____

    _____

27. The _____ tract originates from the red nucleus.

28. The vestibular nucleus gives rise to the _____ tract.

29. The two extrapyramidal tracts involved in the regulation of posture and muscle tone are the _____ and the _____.

30. The _____ tracts conduct impulses for reflexes that change the position of the head, eyes, or neck in response to auditory or visual stimuli.

31. The _____ nuclei are the most important and complex components in the extrapyramidal system.

32. The cerebral nuclei do not _____ a movement, but provide the general pattern and rhythm for the movement once it has been initiated by the motor cortex.

33. The pattern and rhythm of walking seems to be mediated by the _____ nucleus and the _____.

34. The cerebellum receives proprioceptive information from the spinal cord along the _____ tracts.

35. The cerebellum also receives sensory information from the reticular formation and the _____ nucleus.

36. In addition to overseeing the postural muscles of the body, the cerebellum also provides programming and tuning to _____ and _____ movements.

37. The cerebellum can coordinate muscle contractions for complex muscles by _____ how long muscle contractions will take.

38. Normal movement involves a cooperative effort between the _____ and _____ systems in conjunction with the _____.

39. Parkinson's disease is due to a loss of the neurotransmitter _____ in the cerebral nuclei.

40. Write the name of the brain wave in front of the statement which it best matches.

_____ have a large amplitude and low frequency

_____ dominant wave in normal resting adults with their eyes closed

_____ high frequency waves characteristic of adults who are con centrating on a task

_____ found in children and highly frustrated adults

41. Active dreaming occurs during _____ sleep.

42. The majority of a night's sleep is spent in _____ sleep.

43. The initial periods of REM sleep last about _____ minutes.

44. REM periods alternate with _____ sleep.

45. The part of the body that seems to need sleep is the _____.

46. Arousal from sleep is brought about by the _____.

47. The neurotransmitter _____ is associated with the maintenance of RAS activity while the neurotransmitter _____ depresses RAS activity and promotes sleep.

48. The type of memory that is used when looking up a telephone number is _____.

49. _____ are part of our general consciousness.

50. Short term memory is probably due to _____ circuits.

51. A facilitated memory circuit is termed a memory _____.

52. The _____ system plays a key role in the conversion of short term to long term memory.

53. _____ involves the integration of long term memories and their use to direct or modify motor behaviors.

54. In _____ amnesia an individual loses memories of past events.

55. In _____ amnesia a person can recall past events but cannot store additional memories.

56. A person can hear words but cannot interpret their meaning. The person probably has damage to the general _____ area.

57. The area which is responsible for abstract intellectual functions is the _____.

58. The _____ hemisphere possesses the general interpretive and speech centers.

59. Nonverbal imagery resides in the _____ hemisphere.

60. The _____ hemisphere is dominant in most people.

61. A person solving a math problem would be using his/her _____ hemisphere.

62. Match the clinical condition listed below with the appropriate descriptive statement which follows.

a. focal seizure   b. grand mal   c. petit mal   d. Down's syndrome   e. Huntington's disease   f. Alzheimer's disease

_____ an inherited disease which first appears in adults as a progressive deterioration of mental abilities

_____ non-motor epilepsy in which seizures last less than ten seconds

_____ type of senility that sets in between the ages of 50 and 60

_____ attack is characterized by powerful, uncoordinated muscular contractions of the face, eyes, and limbs

_____ caused the possession of an extra copy of chromosome 21

_____ general term for a temporary display of abnormal, involuntary motor patterns

SELF TEST

1. A major difference in integration carried out by the brain as compared to the spinal cord is the

   a. greater number of sensory neurons in the brain. b. greater number of motor neurons in the brain. c. faster speed at which the brain operates. d. greater number and diversity of neurotransmitters and neuromodulators which are possessed by the brain. e. more than one of the above is correct

2. The correct sequence of events for neural processing both in the brain and in the cord is

   a. input - convergence - processing - divergence - output b. input - divergence - convergence - processing - output c. input - processing - convergence - divergence - output d. input - divergence - processing - convergence - output e. output - divergence - processing - convergence - input

3. Which of the following is a neurotransmitter in the brain?

   a. acetylcholine b. norepinephrine c. serotonin d. dopamine e. more than one of the above is correct.

4. A neuromodulator which reduces pain is

   a. acetylcholine. b. serotonin. c. endorphin. d. dopamine. e. glutamic acid.

5. The ascending tract which terminates at the nucleus gracilis is the

   a. fasciculus cuneatus. b. fasciculus gracilis. c. anterior spinothalamic tract. d. posterior spinothalamic tract. e. rubrospinal tract.

6. Which of the following is a motor tract for voluntary muscle control?

   a. reticulospinal b. tectospinal c. rubrospinal d. lateral corticospinal e. anterior spinothalamic

7. A pain in the left arm would be transmitted to the brain by the

   a. left lateral spinothalamic tract. b. right lateral spinothalamic tract. c. right reticulospinal tract. d. left reticulospinal tract. e. fasciculus cuneatus.

8. The extrapyramidal tracts include the

   a. rubrospinal. b. reticulospinal. c. vestibulospinal. d. tectospinal. e. more than one of the above is correct

9. The brain region which is responsible for the pattern and rhythm of movement is the

   a. motor cortex. b. pyramid. c. cerebral nuclei. d. cerebellum. e. pons.

10. The cerebellum

    a. receives proprioceptive information from the spinal cord. b. receives sensory information from the reticular formation. c. oversees the postural muscles of the body. d. provides programming and tuning for all voluntary and involuntary movement. e. all of the above are correct

11. The brain wave which is characteristic of a normal adult with eyes closed is the

    a. alpha wave. b. beta wave. c. delta wave. d. theta wave. e. gamma wave.

12. Dreaming occurs during

    a. deep sleep. b. slow wave sleep.   c. fast wave sleep.   d. REM sleep.   e. more than one of the above is correct.

13. Waking up from sleep is brought about by the

    a. hypothalamus.   b. pons.   c. cerebellum.   d. RAS   e. REM.

14. Short term memory is due to

    a. new synapses.   b. synthesis of a memory molecule.   c. reverberating circuits. d. dopamine synthesis.   e. acetylcholine suppression.

15. An inherited disease that has its first symptoms in young adults is

    a. Down's syndrome.   b. cerebral palsy.   c. Tay-Sach's disease   d. Huntington's disease.   e. petit mal epilepsy.

CHAPTER 16

The Nervous System:
Autonomic Divisions

OVERVIEW

One of the most important functions of the nervous system is the regulation of the internal nonvoluntary organs. These are the organs that are responsible for maintaining stable levels of fluids, electrolytes, nutrients, hydrogen ion and other elements essential for the proper functioning of the cells. In other words, these are the organs that are responsible for homeostasis. The principal branch of the nervous system that is responsible for this regulation is the autonomic division, and in this chapter you will examine its anatomy and physiology.

The autonomic division consists of two major subdivisions; the sympathetic and parasympathetic divisions. Most organs receive innervation from both divisions, and when dual innervation does occur, normally one division increases the activity of the target organ while the opposite division slows its activity down.

The autonomic division is usually considered to be a subdivision of the peripheral nervous system and is often said to be strictly a motor division. Both of these ideas are not completely correct. First of all the preganglionic fibers of the ANS arise in the brain and cord which are part of the CNS. Second, while it is true that the motor fibers of the ANS for discrete anatomical structures, sensory fibers do exist, they are not anatomically distinct as they are contained in nerves along with somatic sensory fibers.

As you proceed through this chapter pay particular attention to the effects that each division has on the respective target organs.

CHAPTER OUTLINE

   A. Introduction
   B. Gross anatomy and organization

    1. The sympathetic division
    2. The parasympathetic division
C. Anatomical and functional relationships
D. Physiology and function
    1. The sympathetic division
    2. The parasympathetic division
E. Control and integration
    1. Visceral reflexes
    2. Biofeedback

LEARNING ACTIVITIES

1. The ANS provides continuous and automatic control over _____ activities.

2. In the autonomic divisions there is always a _____ imposed between the CNS and the peripheral effector.

3. Visceral motor neurons in the CNS are known as _____ neurons.

4. The _____ visceral neurons arise in the autonomic ganglia.

5. First order neurons are also termed _____ fibers while second order neurons are termed _____ fibers.

6. Because preganglionic fibers of the sympathetic division arise from the thoracic and lumbar regions of the spinal cord it is sometimes refered to as the _____ division.

7. Because preganglionic fibers of the parasympathetic division arise from the brain and sacral division of the spinal cord it is sometimes called the _____ division.

8. The sympathetic output emerges between levels T1 and _____ of the spinal cord.

9. Preganglionic fibers of the sympathetic division exit the spinal nerve via the _____ ramus.

10. The sympathetic ganglia that form a chain along the sides of the spinal cord are called the _____ ganglia.

11. The postganglionic fibers from the sympathetic ganglia return to the spinal nerve via the _____ ramus.

12. Postganglionic sympathetic fibers that have to reach target organs not supplied by the spinal nerves exit the sympathetic ganglia by means of _____ nerves.

13. There is considerable divergence in the ANS and the ratio of preganglionic fibers to postganglionic fibers can reach _____.

14. Label figure 16.1, the sympathetic division, using the terms listed below. Refer to figures 16.2 and 16.3 of your textbook for assistance.

gray ramus	white ramus	spinal nerve	posterior root
ganglion	visercal sensory neuron		preganglionic fiber
spinal cord	autonomic (splanchnic) nerve		sympathetic
ganglion	postganglionic fiber		

15. The abdominopelvic viscera receive sympathetic postganglionic fibers from the _____ ganglia.

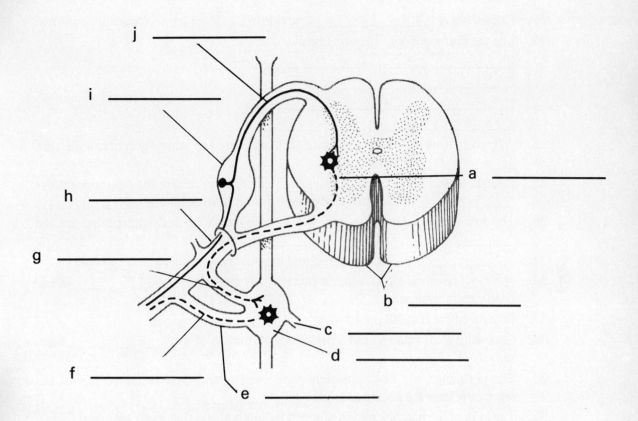

j _____

i _____

h _____

g _____

f _____

e _____

a _____

b _____

c _____

d _____

Figure 16.1

16. _____ nerves are formed from sympathetic preganglionic fibers that skip the sympathetic chain ganglia.

17. The splanchnic nerves from the thoracic regions synapse with the _____ ganglion, _____ ganglion, and adrenal medulla.

18. The _____ innervates the stomach, liver, pancreas, and spleen.

19. The _____ innervates the small intestine and the initial segment of the large intestine.

20. The splanchnic nerves from the lumbar region synapses with the _____ ganglion.

21. The _____ ganglion provides innervation to the large intestine, kidney, bladder, and sex organs.

22. As a general rule the preganglionic fibers of the sympathetic division are _____ and the postganglionic fibers are _____.

23. In the parasympathetic division, preganglionic fibers are _____ and postganglionic fibers are _____.

24. There is a minimum of _____ in the parasympathetic division.

25. The cranial parasympathetic outflow travels over the _____, _____, _____, and _____ cranial nerves.

26. Cranial nerve _____ provides 75% of the parasympathetic outflow.

27. List the four autonomic plexuses below.

_____

_____

_____

_____

28. All autonomic preganglionic fibers release the neurotransmitter _____.

29. Postganglionic fibers of the parasympathetic division release the neurotransmitter _____ while those of the sympathetic division release _____.

30. The two receptor types that respond to norepinephrine and epinephrine are the _____ and _____ receptors.

31. _____ receptors respond to both norepinephrine and epinephrine.

32. The most common alpha receptor functions by causing _____ of excitable cell membranes.

33. Beta receptors respond primarily to _____.

34. Once activated, beta receptors trigger the formation of a _____ messenger.

35. The postganglionic fibers that innervate sweat glands and blood vessels of the muscles release the neurotransmitter _____.

36. The bulk of the norepinephrine released by sympathetic fibers is reabsorbed for use again and the remainder is broken down by _____ and _____.

37. The _____ receptor for acetylcholine is found on the membrane of the parasympathetic ganglion cell.

38. The target cells for the postganglionic fibers of the parasympathetic are the _____ type and may produce excitation or inhibition.

39. _____ blockers eliminates peripheral vasoconstriction.

40. _____ blockers decrease heart rate and force of contraction.

41. During "rest and repose" the body is dominated by the _____ division.

42. During "fight or flight" situations the body is dominated by the _____ division.

43. For the statements listed below place a "P" in the space provided if the statement applies to the parasympathetic division and a "S" if it applies to the sympathetic division. Refer to table 16.2 in your textbook for assistance.

_____ initiates erection of the male sex organ

_____ stimulation causes a slow down in heart rate

_____ increased activity results in dilation of the digestive tract sphincters

_____ stimulation from this division reduces urine production

_____ stimulates the adrenal medulla to secrete

_____ sends no innervation to the sweat glands

_____ increased activity results in pupil dilation

_____ increases the diameter of the respiratory passageways

_____ causes vasoconstriction of the blood vessels of the skin

_____ causes vasodilation of the blood vessels to the muscles

_____ causes constriction of the pupils

44. _____ reflexes are the basic functional units of the ANS.

45. The background activity of the ANS, even in the absence of stimulation, is termed autonomic _____.

46. Autonomic tone normally keeps the peripheral blood vessels at about _____ of their maximum diameter.

47. Autonomic tone is controlled by _____ in the brain.

48. The overall regulation of the ANS occurs mainly in the _____.

49. Although the autonomic nervous system is not under conscious control, it is possible to master limited voluntary control through _____ techniques.

SELF TEST

Circle the correct answer to each question.

1. The parasympathetic division is given the anatomical name of the _____ division.

a. thoracolumbar   b. craniosacral   c. thoracocranial   d. thoracosacral   e. lumbosacral

2. The preganglionic fiber of a sympathetic pathway synapses in one of the chain ganglia or a

a. splanchnic nerve.   b. brain nuclei.   c. collateral ganglion.   d. target organ   e. smooth muscle.

3. Preganglionic fibers of the sympathetic division exit the spinal nerves via the

a. gray ramus.   b. ventral ramus.   c. dorsal ramus.   d. white ramus   e. peripheral ramus.

4. Splanchnic nerves connect chain ganglia with

a. the spinal cord.   b. the brain.   c. target tissues.   d. collateral ganglia.   e. none of the above.

5. The ganglia of the parasympathetic division are usually located

a. in chains along the spinal cord.   b. in clusters behind the cord.   c. intramurally or near the target organ.   d. associated with the aorta and its major branches.   e. none of the above.

6. All preganglionic fibers of the ANS release

a. norepinephrine.   b. epinephrine.   c. acetylcholine.   d. serotonin.   e. dopamine.

7. Norepinephrine that is not reabsorbed by the nerve fiber may be broken down by the enzyme.

a. MAO.   b. acetylcholinesterase.   c. norepinephrinase   d. NAD.   e. more than one of the above is correct

8. Both epinephrine and norepinephrine can activate _____ receptors.

a. muscarinic   b. nicotinic   c. alpha   d. beta   e. gamma

9. When activated, beta receptors alter cellular metabolism by initiating

a. depolarization.   b. synthesis of a second messenger.   c. epinephrine synthesis.   d. contraction.   e. acetylcholine destruction.

10. The receptors found on the target tissue of the parasympathetic division are the
_____ type.
a. alpha   b. beta.   c. gamma.   d. nicotinic.   e. muscarinic.

11. The nerve network that regulates the heart is the

a. pulmonary plexus.   b. cardiac plexus.   c. hypogastric plexus.   d. brachial plexus.   e. cervical plexus.

12. A person who was suffering from high blood pressure would most likely take a (an)

a. beta-blocker.   b. muscarine-blocker.   c. nicotine-blocker.   d. alpha-blocker. e. gamma-blocker.

13. A person participating in a football game would be

a. dominated by the sympathetic division.   b. dominated by the parasympathetic division.   c. in essentially the same autonomic state as during rest.   d. dominated by either the sympathetic or parasympathetic depending upon the situation.   e. none of the above.

14. Increased sympathetic stimulation will cause

a. constriction of the pupil.   b. increased digestive activity.   c. increased urine production.   d. increased secretion of digestive glands.   e. increased perspiration.

15. Even with no stimuli to the ANS, most organs innervated by it will still be affected because of

a. biofeedback.   b. autonomic tone.   c. visceral reflexes.   d. Horner's syndrome. e. none of the above.

CHAPTER 17
Sensory Function

OVERVIEW

All motor activity directed by the nervous system is the result of received stimuli. According to the nature of the stimulus, integration will direct a meaningful motor response. The source of these stimuli are the sensory receptors of the nervous system. It is these receptors that respond to changes in both the internal and external environment, alerting the integrating centers of a possible need for action. These sensory receptors are the subject of this chapter which is the final one in your tour of the nervous system.

The chapter begins with a consideration of the nature of sensory receptors, what they are, and how they function, followed by a consideration of the various types of receptors. You will first be introduced to the general senses. These are the senses that are associated with the skin such as touch, temperature, pressure, and pain. Included in the general senses are also those of proprioception and chemical detection. Next comes the special senses, taste, smell, hearing, equilibrium, and vision. The receptors for these senses are located primarily in the head.

Keep in mind that regardless of the kind of receptor or the type of stimulus which it responds to, all receptors ultimately convert stimuli into action potentials which are all alike. It is the region of the central nervous system to which a given action potential is directed that will determine whether the sensation is perceived as vision, hearing, pain, or some other sensation.

CHAPTER OUTLINE

A. Introduction
B. Principles of receptor function
    1. The classification of receptors
    2. Receptor physiology and sensory coding

    3. Central processing and adaptation
- C. The general senses
  - 1. Nociception
  - 2. Chemoreception
  - 3. Thermoreception
  - 4. Mechanoreception
    - a. Tactile receptors
    - b. Baroreception
    - c. Proprioreception
- D. The special senses
  - 1. Olfaction
  - 2. Gustation
  - 3. Equilibrium and hearing
    - a. General anatomy and organization
      - (1) The external and middle ear
      - (2) The inner ear
      - (3) Receptor function
    - b. The vestibular sense organs
      - (1) The semicircular canals
      - (2) The utricle and saccule
      - (3) Central processing
    - c. The cochlea
      - (1) Principles of sound generation and perception
      - (2) Cochlear structure and function
      - (3) Central processing
  - 4. Vision
    - a. Accessory structures of the eye
    - b. Anatomy of the eye
      - (1) The fibrous tunic
      - (2) The vascular tunic
      - (3) Aqueous humor
      - (4) The vitreous chamber and the neural tunic
      - (5) The lens
    - c. Light and photoreception
      - (1) Retinal organization and receptor function
      - (2) Visual responses to changing light levels.
      - (3) Color vision
    - d. Retinal processing
    - e. Central processing of visual information
- E. Clinical patterns

LEARNING ACTIVITIES

Complete the following items by supplying the appropriate word or phrase.

1. The first, and simplest type of sensory receptor is an unspecialized neuron known as a free _____ ending.

2. In the second type of receptor, the dendritic end of a sensory receptor surrounded by Schwann cells is then called an _____ nerve ending.

3. A third pattern of receptor utilizes sensory neurons to contact special
   _____ cells that act as the receptor.

4. The fourth and most specialized type of receptor are complex organs that contain
   _____ neurons as sensory receptors.

5. Sensations can be categorized as _____ or _____.

6. Receptors for the _____ senses are scattered throughout the body
   while those for the special senses are found in restricted areas.

7. A functional classification of receptors based upon the environments to which they
   respond would include the following.

   _____ respond to the external environment.

   _____ respond to the internal environment.

   _____ track the position of muscles and joints.

8. Using the receptor classification based upon the nature of the stimulus, write the
   receptor type in front of the statement which best describes that receptor.

   _____ respond to mechanical distortion.

   _____ monitor the chemical composition of the body.

   _____ are sensitive to light.

   _____ respond to any type of stimulus that produces pain.

   _____ respond to changes in temperature.

9. There are three basic steps that must occur in all receptors in order to activate a sensory neuron. These steps are listed below. Arrange them in proper order of occurence
   in the space provided by writing the letters identifying each step in the proper order.

   _____

   a. Generates an action potential on the afferent nerve fiber.

   b. A stimulus physically alters the receptor cell membrane.

   c. There is a change in the permeability of the receptor cell membrane to ions.

10. The brain "knows" that your left big toe has been stimulated because sensory path-
    ways are organized into precise neural pathways known as _____
    lines.

11. The brain "knows" that your left big toe is in pain because the labeled line which is
    activated indicates the nature or _____ of the sensation.

12. The area monitored by a single receptor represents the receptor _____.

13. The strength of a stimulus is interpreted by the brain based upon the
    _____ of the action potentials being transmitted along the labeled
    line.

14. Sensory receptors which are always active and responding to background level of
    stimuli are _____ while those which are not active until the environ-
    ment changes are _____.

15. The conscious awareness of a stimulus is known as _____.

16. Most sensations are processed in _____ along the cord or brain stem
    and do not result in perception.

17. In _____ adaptation, a sensory receptor stops responding to a con-
    stant level stimulus.

18. In _____ adaptation, nuclei of the CNS can either inhibit or facilitate
    sensitivity to stimuli.

19. Nociceptors (pain receptors) consist of _____ endings.

20. Fast or prickling pain sensations are carried by _____ fibers while slow or aching pain is carried by _____ fibers.

21. Deep pain that is perceived as arising on the surface of the body is known as _____ pain.

22. Other than olfaction and gustation, chemoreceptors in the body respond to _____ ion and _____.

23. Temperature receptors are scattered immediately beneath the surface of the _____.

24. Thermoreceptors represent another population of _____ nerve endings.

25. Thermoreceptors are most active when the temperature is _____.

26. The three categories of mechanoreceptors are _____, _____, and _____.

27. Match the tactile receptor listed below with the appropriate descriptive statement which follows.

a. root hair plexus   b. Merkel's discs   c. Meissner's corpuscles   d. Pacinian corpuscles   e. Ruffini corpuscles   f. free nerve ending

_____ only receptors found on the cornea of the eye

_____ primarily monitors distortions and movements across the body surface

_____ encapsulated receptors that respond to changing pressure

_____ pressure receptors which are chronically active

_____ fine touch receptors located in the dermis of the skin

_____ fine touch receptors that make contact with cells in the stratum germinativum

28. Baroreceptors monitor changes in _____.

29. Baroreceptors consists of free nerved ending found in _____ tissues of distensible organs.

30. The _____ organ is responsible for the tendon reflex.

31. The _____ becomes active when a muscle begins to relax.

32. The _____ organs consist of small areas of mucous epithelium on either side of the nasal septum.

33. _____ glands produce a pigmented mucous that covers the epithelium of the olfactory epithelium.

34. Airborne molecules dissolve in the _____ that covers the olfactory epithelium.

35. The olfactory receptors are highly specialized _____ that monitor the chemical contents of the mucous covering the olfactory epithelium.

36. As few as _____ molecules can activate an olfactory receptor.

37. Axon bundles from the olfactory organs enter the olfactory _____ of the cerebrum.

38. Olfaction is perceived by the _____ cortex.

39. Extensive olfactory fibers to the _____ system and _____ explain the strong emotional and behavioral effects that can be produced by certain smells.

40. The gustatory (taste) receptors are clustered into taste _____.

41. Each gustatory cell contains a slender microvillus termed a _____ hair which samples the chemicals dissolved in saliva.

42. The taste buds on the tongue lie along epithelial projections termed
    _____.

43. The three types of papillae found on the human tongue are the
    _____, _____, and _____.

44. The four primary taste sensations are _____, _____,
    _____, and _____.

45. The taste receptors for _____ and _____ are much
    more sensitive than are those for _____ or _____.

46. The taste buds are monitored by the _____, _____, and
    _____ cranial nerves.

47. The sensory afferent fibers from the taste buds synapse with the nucleus
    _____ in the medulla.

48. The postsynaptic axons from the taste buds enter the medial _____
    and then synapse in the thalamus from where fibers are directed to the appropriate
    regions of the primary sensory cortex.

49. The perception of taste involves not only the taste receptors, but general receptors
    such as touch and especially the _____ receptors.

50. Label figure 17.1, gustatory pathways, using the terms listed below. Color each struc-
    ture. Refer to figure 17.8 in your textbook for assistance.

sweet	salt	sour	bitter	primary sensory cortex
thalamic nucleus		medial lemniscus		solitary nucleus
cranial nerve IX		cranial nerve VII		cranial nerve X

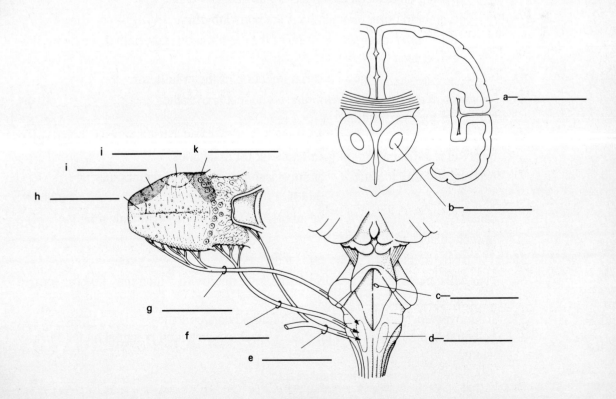

Figure 17.1

51. Match the parts of the ear listed below with the appropriate descriptive statement which follows.

a. external ear   b. pinna   c. external auditory canal   d. ceruminous glands   e. tympanic membrane   f. pharyngotympanic tube   g. auditory ossicles   h. inner ear   i. membranous labyrinth   j. osseous labyrinth   k. endolymph   l. perilymph   m. vestibule   n. utricle   o. cochlea   p. semicircular canals   q. cochlear duct   r. round window   s. oval window

_____ division of the bony labyrinth that houses the receptors for position and balance

_____ membranous window which is bound to the auditory ossicles

_____ region of the ear that begins with the oval window

_____ flap that surrounds the external auditory meatus

_____ ear region that ends at the tympanic membrane

_____ connect the tympanic membrane to the inner ear

_____ one of the membranous sacs found inside of the bony vestibule

_____ fluid contained inside of the membranous labyrinth.

_____ membranous window that is not attached to the auditory ossicles

_____ region of the bony labyrinth that houses the organ of hearing

_____ fluid that separates the membranous and bony labyrinths

_____ structures that contain receptors that detect turning motion

_____ another name for this structure is the eardrum

_____ structure that permits the middle ear to communicate with the nasopharynx

_____ glands which secrete a waxy material in the external ear

_____ ear region that contains the external auditory canal

_____ membrane structure that lines the bony labyrinth

_____ the slender, elongated portion of the membranous labyrinth that extends into the cochlea

52. _____ is a bacterial infection of the middle ear.

53. Vibration of the tympanic membrane is conducted over the _____ to the surface of the oval window.

54. Label figure 17.2, structure of the ear, using the terms listed below. Color each structure. Refer to figure 17.9 in your textbook for assistance.

external auditory meatus   auditory canal   tympanic membrane
acoustic nerve   middle ear   inner ear   cochlea
canals   vestibule   pharyngotympanic tube
semicircular

55. The basic receptor cell of the ear is the _____ cell.

56. Hair cells are so named because of the long microvilli which they possess termed

_____.

57. It is movement of the stereocilia which causes changes in the _____ potential of the hair cell and ultimately results in an action potential on the sensory neuron.

58. There are three semicircular canals, the _____, the _____, and the _____ canal.

59. The expanded base of each canal, the _____, is where the receptor cells are located.

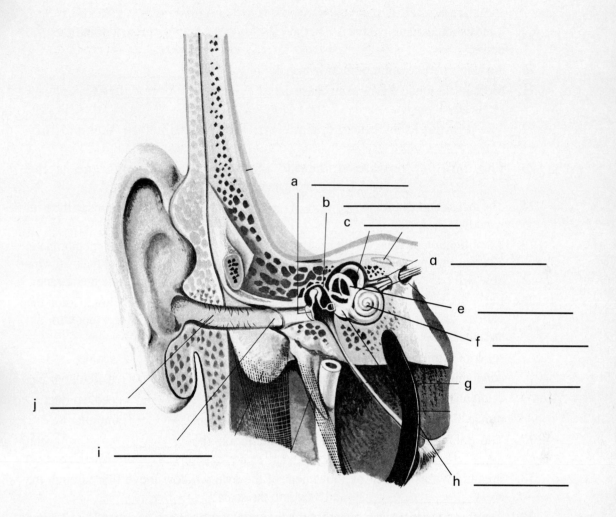

a _____

b _____

c _____

a _____

e _____

f _____

g _____

j _____

i _____

h _____

Figure 17.2

60. The stereocilia of the hair cells in the ampulla are embedded in a gelatinous structure known as the _____.

61. Movement of _____ puts pressure on the cupula, distorting the stereocilia and generating receptor potentials.

62. The semicircular canals open into the _____.

63. The function of the _____ is to return excess endolymph to the circulation.

64. The utricle and saccule detect gravity and linear _____.

65. The hair cells of the utricle and saccule are contained in the _____.

66. The receptor cells in the maculae lie under a mass of crystals termed _____.

67. Receptor potentials are produced by the hair cells of the maculae in response to pressure exerted by the _____.

68. The sensory neurons of the vestibule are located within a pair of _____ ganglia.

69. The vestibular ganglia give rise to the _____ branch of the acoustic nerve.

70. An inappropriate sense of motion best defines _____.

71. Mechanical sound waves in air cause the _____ to vibrate at their frequency.

72. The tympanic membrane transfers the vibrations to the auditory _____.

73. The auditory ossicle connected to the tympanic membrane is the _____, which in turn connects to the _____.

74. The incus connects to the _____ which in turn is attached to the oval window.

75. Amplification of weak auditory signals occurs because the tympanic membrane is 22 times the area of the oval window. This means that the deflection of the oval window will be _____ times as great at that of the tympanic membrane.

76. Label 17.3, the membranous labyrinth, using the terms listed below. Color the various structures. Refer to figures 17.9, 17.10, and 17.13 in your textbook for assistance.

superior semicircular canal   posterior semicircular canal   saccule
endolymphatic sac   endolymphatic duct   vestibular branch
vestibular ganglia   cochlear nerve   cochlear duct
utricle   lateral semicircular canal   ampulla

77. The cochlea is divided into three chambers, the _____, the _____, and the _____.

78. Pressure waves created by movement of the oval window move first through the scala _____ and then into the scala _____.

79. The pressure wave that moves through the endolymph of cochlea expends itself upon the _____.

80. The organ of hearing is the organ of _____.

81. The organ of Corti lies on the _____ membrane which separates the scala tympani from the scala media.

82. The stereocilia of the hair cells in the organ of Corti are in contact with the _____ membrane which lies above them.

83. Pressure waves moving through the scala tympani distort the basilar membrane forcing the stereocilia against the _____ membrane.

84. Bending of the stereocilia generate _____ potentials.

85. High frequency sounds cause vibration of the basilar membrane close to the _____ while low frequency sounds cause vibration towards the apex or end of the basilar membrane.

86. _____ deafness results from conditions in the middle ear that block the normal transfer of vibrations.

87. In _____ deafness the problem is in the cochlea or along the auditory pathway.

l _____

k _____

j _____

i _____

a _____

b _____

c _____

d _____

e _____

f _____

h _____

g _____

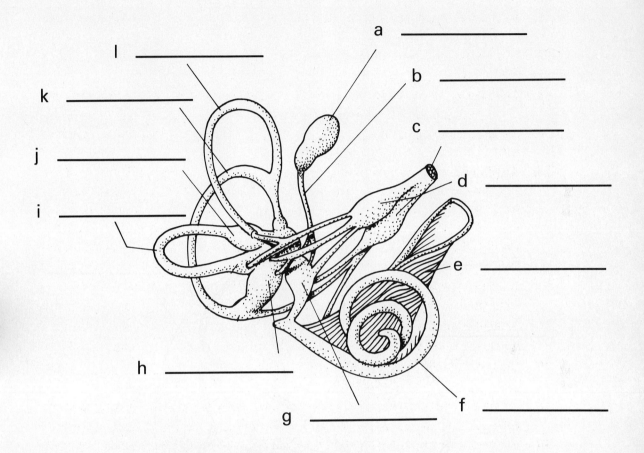

Figure 17.3

88. The sensory neurons of the _____ ganglion are housed within the cochlea.

89. The afferent fibers of the spinal ganglion form the _____ branch of the acoustic nerve.

90. Complete the sequence of synapses for the fiber of the cochlear branch.
    cochlear nucleus (medulla) - _____ - medial geniculate - _____.

91. The auditory cortex can discriminate sound frequencies because it has a map of the _____, and knows which portion is being stimulated.

92. The cortex also determines the _____ that a sound came from.

93. Label figure 17.4, the cochlea, using the terms listed below. Color each structure. Refer to figure 17.6 in your textbook for assistance.

cochlea   scala vestibuli   tectorial membrane
scala media (cochlear duct)   basilar membrane   organ of Corti
spiral ganglion   scala tympani

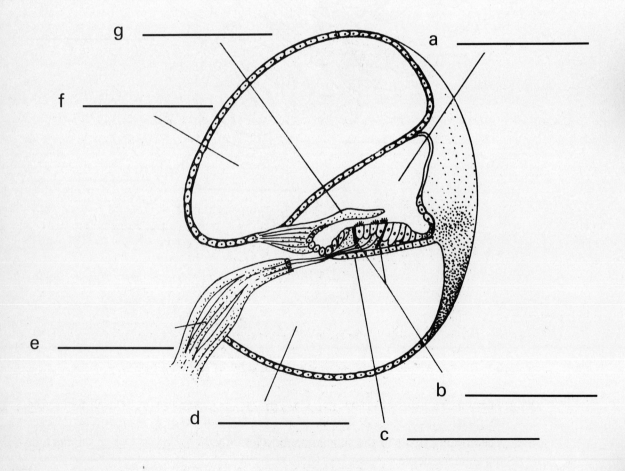

g _____   a _____

f _____

e _____

d _____   b _____
                           c _____

Figure 17.4

94. The eyelids, or palpebrae are connected at the medial _____ and the lateral _____.

95. The membrane that lines the eyelids and covers the exposed portion of the eye is the _____.

96. Tears are produced by the _____ gland.

97. Label figure 17.5, lacrimal appartus, with the terms listed below. Refer to figure 17.18 in your textbook for assistance.

superior lacrimal gland	inferior lacrimal gland	lateral canthus
lacrimal puncta	lacrimal canals	nasolacrimal duct

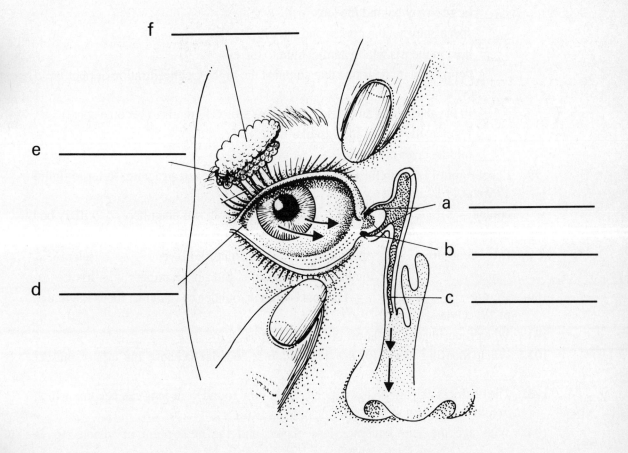

Figure 17.5

98. Match the terms listed below with the appropriate descriptive statements which follow.

a. fibrous tunic b. vascular tunic c. neural tunic d. sclera e. cornea f. iris g. ciliary body h. choroid i. pupil j. anterior chamber k. suspensory ligaments l. aqueous humor m. canal of Schlemm n. retina o. vitreous body p. vitreous chamber q. optic disc r. lens s. fovea centralis t. posterior chamber

_____ blind spot on the retina

_____ attach the lens to the inner margin of the ciliary body

_____ the outer layer of the eye

_____ transparent part of the fibrous tunic

_____ structures that drain aqueous humor

_____ gelatinous substance found in the vitreous chamber

_____ contains the photoreceptors

_____ tunic that contains the choroid layer

_____ structure that controls the focus of the eyes

_____ fluid that maintains proper ocular pressure and supplies nutrients to avascular eye tissues

_____ the space between the suspensory ligaments and the iris

_____ structure that surrounds the pupil

_____ large cavity behind the lens

_____ space between the cornea and the iris

_____ hole in the iris which admits light to the eye

_____ depression in the retina that contains the highest concentration of photoreceptors

_____ in the absence of an outside force, this structure tends to become spherical

_____ pigmented layer of the vascular tunic

_____ middle layer of the eye

99. Label figure 17.6 with the terms listed below. Color each structure. Refer to figure 17.9 and 17.20 in your textbook for assistance.

cornea	conjunctiva	posterior chamber	ciliary body
suspensory ligament	lateral rectus muscle	vitreous body	pupil
fovea centralis	optic nerve	choroid layer	sclera
retina	medial rectus muscle	anterior chamber	iris

100. During _____ the lens becomes rounder or flatter to focus the image on the retina.

101. When focusing at a distance, the lens becomes _____.

102. When focusing close up, the lens becomes rounder because the ciliary muscle _____.

103. The nearness of your vision depends upon how round your lens can become which in turn depends upon the lens' _____.

104. With age the lens becomes less elastic and the near point of vision moves _____ away.

105. Visual _____ which is the clarity of vision, is usually measured with an eye chart.

106. A person with normal visual acuity would have 20/_____ vision.

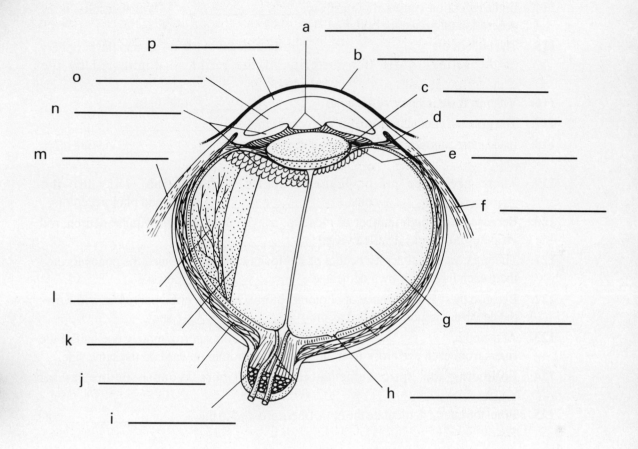

Figure 17.6

107. A person with 20/200 vision would have to stand 20 feet away from an object to see the same detail that a person with normal vision could see at _____ from the object.

108. A _____ represents an energy packet of visible light.

109. The two types of receptor cells in the retina are _____ and _____.

110. Rods are used for _____ vision and cones for _____ vision.

111. The population of cones in the retina contain three types based upon their sensitivity to light colors. These are _____, _____, and _____ cones.

112.   The principal visual pigment in the rods and cones is _____.

113.   The pigment portion of rhodopsin is _____ which is derived from vitamin A.

114.   Light reception causes rhodopsin to _____ which results in the generation of a receptor potential.

115.   The bulk of the _____ are distributed in a band around the periphery of the retina, while the posterior of the retina is dominated by the _____.

116.   Vitamin A deficiency will lead to _____ blindness.

117.   The greatest concentration of cones occurs in the _____.

118.   In the most common form of color blindness, the _____ cones are missing.

119.   There are three major layers of cells in the retina. They are the _____ cells, _____ cells, and photoreceptors.

120.   Because of the high number of rods that converge on a single bipolar neuron, rod vision usually lacks detailed visual _____.

121.   The greatest visual acuity occurs in the fovea centralis because the cones located there each have their own dedicated _____ neuron

122.   Because there is an absence of photoreceptors where the optic nerve leaves the eye, the departure region is known as the _____ spot.

123.   At the optic _____ in the diencephalon, approximately one half of the fibers from each eye cross-over and one half continue onward on the same side.

124.   Following the optic chiasma, the optic fibers synapse in the lateral _____.

125.   From the lateral geniculate the optic fibers proceed to the _____ cortex.

126.   Collateral optic fibers are sent to the superior _____.

127.   Label figure 17.7, regions of the retina, using the terms listed below. Color each layer of cells. Refer to figure 17.26 in your textbook for assistance.

ganglion cell layer   bipolar neuron layer   photoreceptor layer
rod cell   cone cell   optic nerve fibers

SELF TEST

Circle the correct answer to each question.

1.   Receptors which consist of free dendrites constitute the _____ receptor type.

a. encapsulated nerve ending   b. free nerve ending   c. accessory cell   d. modified neuron.   e. more than one of the above is correct

2.   Pain receptors are classified as

a. mechanoreceptors.   b. thermoreceptors.   c. chemoreceptors.   d. photoreceptors.   e. nociceptors.

3.   The intensity of a stimulus can be interpreted based upon

a. the identity of the receptor.   b. the location of the receptor.   c. the receptor field.
d. the frequency of action potentials from the receptor.   e. none of the above.

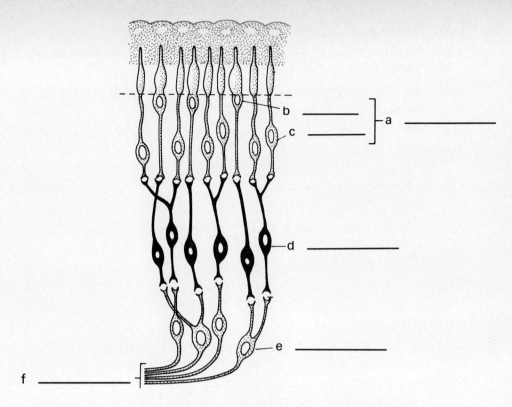

Figure 17.7

4. Perception is a function of the
   a. receptor. b. afferent fibers. c. frequency of action potentials. d. intensity of stimulus. e. interpreting center in the brain.

5. Pain receptors are
   a. free nerve endings. b. encapsulated nerve endings. c. accessory cells.
   d. modified neurons. e. more than one of the above is correct.

6. A tactile receptor would be
   a. Merkel's discs. b. Meissner's corpuscles. c. the root hair plexus. d. Pacinian corpuscles. e. more than one of the above is correct.

7. The receptor of taste is the
   a. olfactory cell. b. gustatory cell. c. basal cell. d. hair cell. e. taste bud.

8. Which of the following is not a primary taste?
   a. sweet b. sour c. bitter d. salt e. peppermint

9. The auditory ossicle that connects to the oval window is the
   a. incus. b. malleus. c. stapes. d. saccule. e. utricle.

10. The organ of Corti is located on the
    a. utricle. b. saccule. c. macula. d. semicircular canal. e. none of the above

11. The receptor cells for sound generate receptor potentials when their stereocilia come in contact with the
    a. basilar membrane. b. tectorial membrane. c. cupula. d. macula e. otoconia.

12. The structure that controls the amount of light entering the eye is the
    a. ciliary body. b. anterior chamber. c. cornea. d. iris. e. lens.

13. The outer covering of the eye is the
    a. choroid. b. sclera. c. retina. d. uvea. e. ciliary body.

14. Focusing of the eye is a function of the

   a. ciliary body.   b. iris.   c. anterior chamber.   d. posterior chamber.   e. retina.

15. The crossing over of optic fibers occurs in the

   a. retina.   b. ganglion layer.   c. optic chiasma.   d. lateral geniculate   e. superior colliculus.

CHAPTER 18
The Endocrine System

OVERVIEW

There is a second command and control system which integrates the body systems. This is the endocrine system, a system of ductless glands that produce metabolic regulators known as hormones. The endocrine system works in concert with the nervous system to control all of the body's functions. When compared to the nervous system we find that the endocrine system affects a greater variety of tissues, has a longer lasting effect, but is slower to respond. Another difference is that the endocrine system has a graded effect on various tissues. By this we mean that the level or concentration of hormones is very important. Recall that the nervous system is largely a digital system where information is coded in the frequency of similar intensity action potentials. Finally, the endocrine system is totally an internal system. Any changes that occur in endocrine functions due to external events are usually mediated by the nervous system. In contrast, a large part of the nervous system is devoted exclusively to monitoring and responding to external events.

The nervous and endocrine systems are the opposite sides of the regulatory coin. Both are interconnected, affecting one another, but the nervous system predominates. Ultimately, virtually all of the endocrine system is under the direct or indirect control of the nervous system.

This chapter begins by introducing you to the nature of hormones and how they exert their effects on target cells. From there you will proceed to a consideration of the interaction between the nervous and endocrine systems to maintain homeostasis. Then begins a survey of the major endocrine glands, the hormones they produce, and the effects of these hormones on target tissues. Finally there is a consideration of hormone interactions including their role in growth, stress, and behavior.

CHAPTER OUTLINE

A. Introduction
B. Hormone structure and function
    1. Mechanism of hormone action
C. Principles of homeostatic control
    1. The hypothalamus and endocrine regulation
D. Endocrine tissues and organs
    1. The pituitary gland
        a. The posterior pituitary
        b. The anterior pituitary
            (1) Hormones of the anterior pituitary
            (2) The hypothalamic control of the anterior pituitary
    2. The thyroid gland and associated endocrine structures
        a. The thyroid gland
            (1) The thyroid follicles
            (2) The C-cells of the thyroid gland
        b. The parathyroid glands
        c. The thymus
    3. The adrenal gland
        a. The adrenal cortex
        b. The adrenal medulla
    4. The endocrine functions of the kidneys and heart
    5. Endocrine tissues of the digestive system
        a. The pancreas
    6. The endocrine tissues of the reproductive system
    7. The pineal gland
E. Patterns of hormonal interaction
    1. Hormones and growth
    2. Hormones and stress
        a. The alarm phase
        b. The resistance phase
        c. The exhaustion phase
    3. Hormones and behavior
F. Clinical patterns

LEARNING ACTIVITIES

Complete each of the following items by supplying the appropriate word or phrase.

1. The effectors of the endocrine system are _____.
2. The three chemical categories of hormones are the _____ derivatives, _____, and _____.
3. The catecholamines include _____ and _____.
4. The hypothalamus, pituitary gland, and pancreas all produce _____.
5. _____ hormones are produced by the adrenal cortex and gonads.
6. Label figure 18.1, the endocrine system, using the terms listed below. Color each
   gland. Refer to figure 18.1 in your textbook for assistance.

   pituitary gland   pineal gland   adrenal gland   thyroid gland
   parathyroid gland   pancreas   teste   ovary

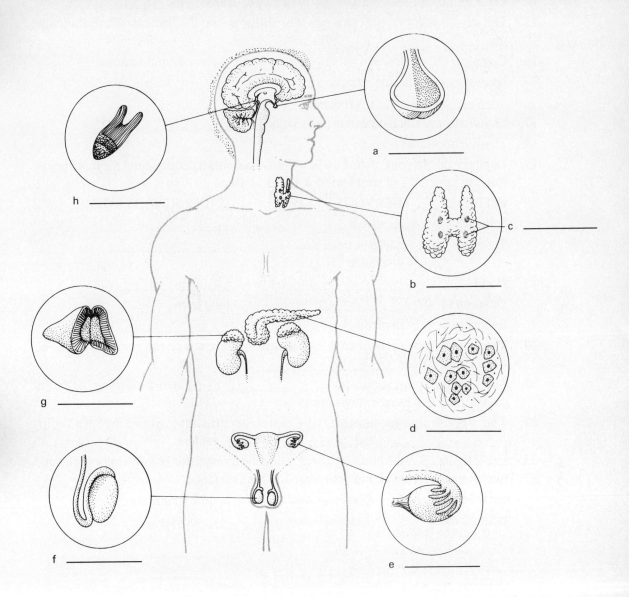

Figure 18.1

7. Hormones alter cellular metabolism by acting upon the cell's _____.

8. The first step in a hormone-target cell interaction is the attachment of the hormone to a cell _____.

9. The hormone receptor complex acts by altering the activity of enzymes present in the cell or by activating _____ in the nucleus.

10. The catecholamines and peptide hormones interact with cell receptors located in the _____.

11. Interaction of the catecholamines and peptide hormones with membrane receptors results in the synthesis of a _____ messenger within the cell.

12. The most important second messenger is _____.

13. Complete the sequence of events presented below for a catecholamine hormone.

    hormone - membrane receptor - _____ activation -

    cyclic AMP (second messenger) - enzyme _____

14. The steroid hormones interact with a receptor located in the _____.

15. The steroid hormone-receptor complex diffuses into the nucleus where it initiates the _____ of mRNA.

16. Complete the sequence of events presented below for a steroid hormone.

    hormone - cytoplasmic receptor - _____ activation -
    _____ synthesis

17. Thyroid hormones bind to receptors in the _____.

18. Thyroid hormones activate _____.

19. Compare the nervous system and endocrine systems by completing the table below. Refer to table 18.1 in your textbook for assistance.

EFFECTS	NERVOUS SYSTEM	ENDOCRINE SYSTEM
Scope	localized	_____
Targets	nervous, muscular, glandular	_____
Onset	_____	_____
Duration	_____	long term
Recovery	immediate	_____

20. The hypothalamus contains autonomic centers that exert neural control over the endocrine cells of the _____.

21. The hypothalamus is also an _____ gland producing ADH and oxytocin as well as other hormones.

22. The hypothalamus controls the anterior pituitary gland by producing _____ and _____ factors.

23. Label figure 18.2, the pituitary gland, using the terms listed below. Color each structure. Refer to figure 18.5 in your textbook for assistance.

    adenohypophysis   pars intermedia   neurohypophysis
    infundibulum   hypothalamus   hypophysis

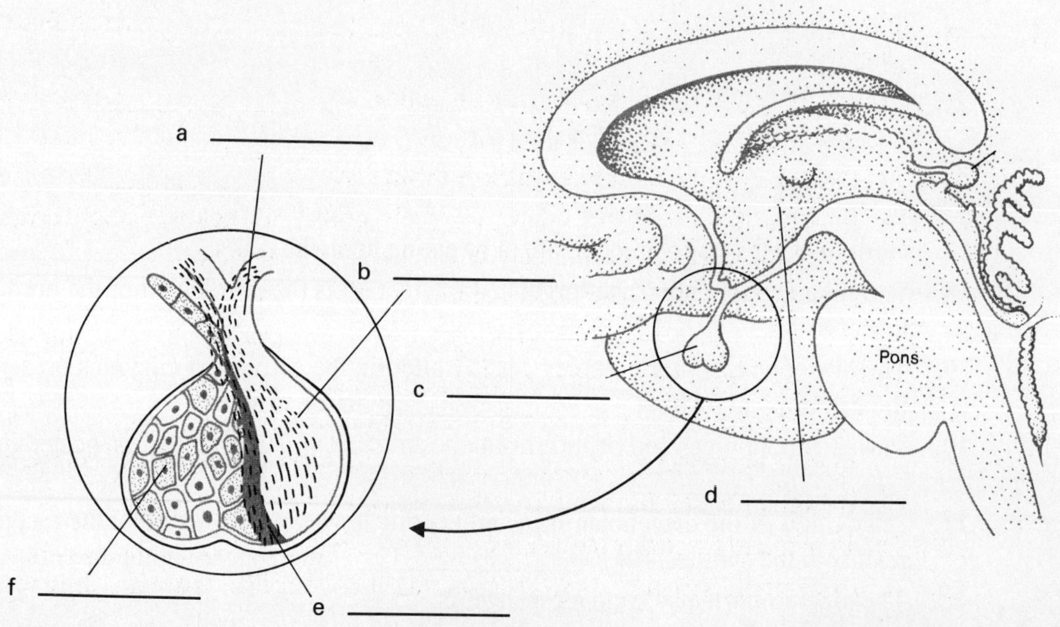

a _____
b _____
c _____
d _____
Pons
f _____
e _____

Figure 18.2

24. An alternate name for the pituitary gland is the _____.

25. The posterior pituitary is also known as the _____.

26. ADH is produced by the _____ nucleus and oxytocin is produced by the _____ nucleus.

27. The primary function of ADH is to decrease the amount of _____ lost from the body.

28. _____ stimulates contraction of smooth muscle in the uterus and the breast.

29. ADH and oxytocin reach the neurohypophysis from the hypothalamus by means of _____ transport along the infundibulum.

30. The anterior pituitary is also known as the _____.

31. Match the adenohypophyseal hormone listed below with the appropriate descriptive statement which follows.

    a. thyroid stimulating hormone (TSH)   b. adrenocorticotrophic hormone (ACTH)
    c. follicle stimulating hormone (FSH)   d. lutenizing hormone (LH)   e. prolactin
    f. growth hormone (GH)   g. melanocyte stimulating hormone (MSH)

    _____ stimulates cell growth and replication by stimulating protein synthesis
    _____ stimulates the thyroid gland
    _____ stimulates the melanocytes of the skin
    _____ in men this hormone stimulates release of the male sex hormones
    _____ promotes the release of steroid hormones from the adrenal cortex
    _____ stimulates the production of milk by the breast
    _____ promotes egg development in women and sperm production in men

32. The releasing and inhibiting factors that control release of the adenohypophyseal hormones reach the adenohypophysis from the hypothalamus via the _____.

33. TSH, ACTH, FSH, and LH are all controlled by _____ factors from the hypothalamus.

34. Prolactin and MSH are regulated by _____ factors from the hypothalamus.

35. Growth hormone has a releasing factor and an _____ factor which controls its concentration.

36. The two lobes of the thyroid are connected by the _____.

37. Label figure 18.3, the thyroid gland, using the terms listed below. Color the various structures. Refer to figure 18.9 in your textbook for assistance.

    left lateral lobe   right lateral lobe   isthmus
    hyoid bone   thyroid cartilage   trachea   follicles

38. Thyroid hormones form in the thyroid _____.

39. All thyroid hormones are derived from the amino acid _____.

40. The thyroid hormone, _____, affects virtually all cells of the body.

41. The overall effect of thyroxine is to increase cell _____.

42. The immediate precursor molecule of thyroxine is _____.

43. In the circulation, thyroxine binds to special proteins known as _____.

44. A second hormone produced by the thyroid which functions in calcium metabolism is _____.

Label: left lateral lobe   right lateral lobe   isthmus   hyoid bone   thyroid cartilage   trachea   follicles

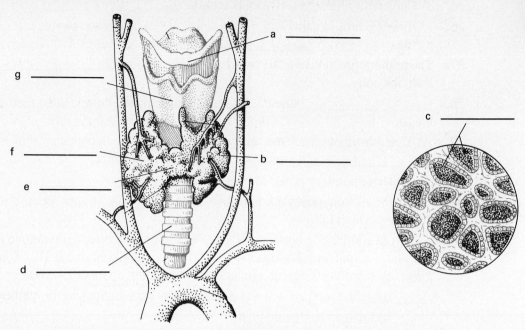

Figure 18.3

45. The overall effect of calcitonin is to _____ serum calcium.

46. It is the _____ cells of the parathyroid glands that produce PTH.

47. The overall effect of PTH is to _____ serum calcium.

48. Write C for calcitonin or P for PTH in front of each statement below which applies to the hormones respectively.

    _____ stimulates osteoclasts

    _____ promotes intestinal absorption of calcium

    _____ stimulates osteoblasts and inhibits osteoclasts

    _____ under its influence the kidney increases its excretion of calcium

    _____ reduces the intestinal absorption of calcium

49. _____, produced by the thymus gland, maintains proper immunological function.

50. Each adrenal gland has two parts, the _____ and the _____.

51. The adrenal cortex is divided into three regions, the zona _____, _____, and _____.

52. The zona reticularis produces small quantities of the _____ sex hormone in both females and males.

53. The zona fasciculata produces the _____.

54. The glucocorticoids increase _____ synthesis and _____ formation.

55. The three most important glucocoritcoids are _____, _____, and _____.

56. In addition to their carbohydrate effects, the glucocorticoids are also anti _____ agents.

57. The zona glomerulosa produces hormones known as _____.

58. The most important mineralocorticoid is _____.

59. Aldosterone increases _____ and water retention.

60. The major stimulus for aldosterone secretion is _____.

61. The adrenal medulla produces the _____.

62. The adrenal medulla is controlled by the _____ division of the ANS.

63. The catecholamines function as _____ messengers.

64. For the statements listed below, write an I in front if catecholamines cause an increase and a D if they result in a decrease.

_____ blood sugar

_____ heart rate

_____ blood pressure

_____ adipose tissue

_____ glycogen

65. Label figure 18.4, the adrenal glands, using the terms listed below. Color each structure. Refer to figure 18.12 in your textbook for assistance.

adrenal gland   capsule   kidney   medulla   cortex

66. The kidney releases the enzyme _____ which converts the blood protein angiotensinogen into angiotensin.

67. The kidney hormone _____ increases the production of red blood cells.

68. List the four major functions of atrial natriuretic factor.

_____

_____

_____

_____

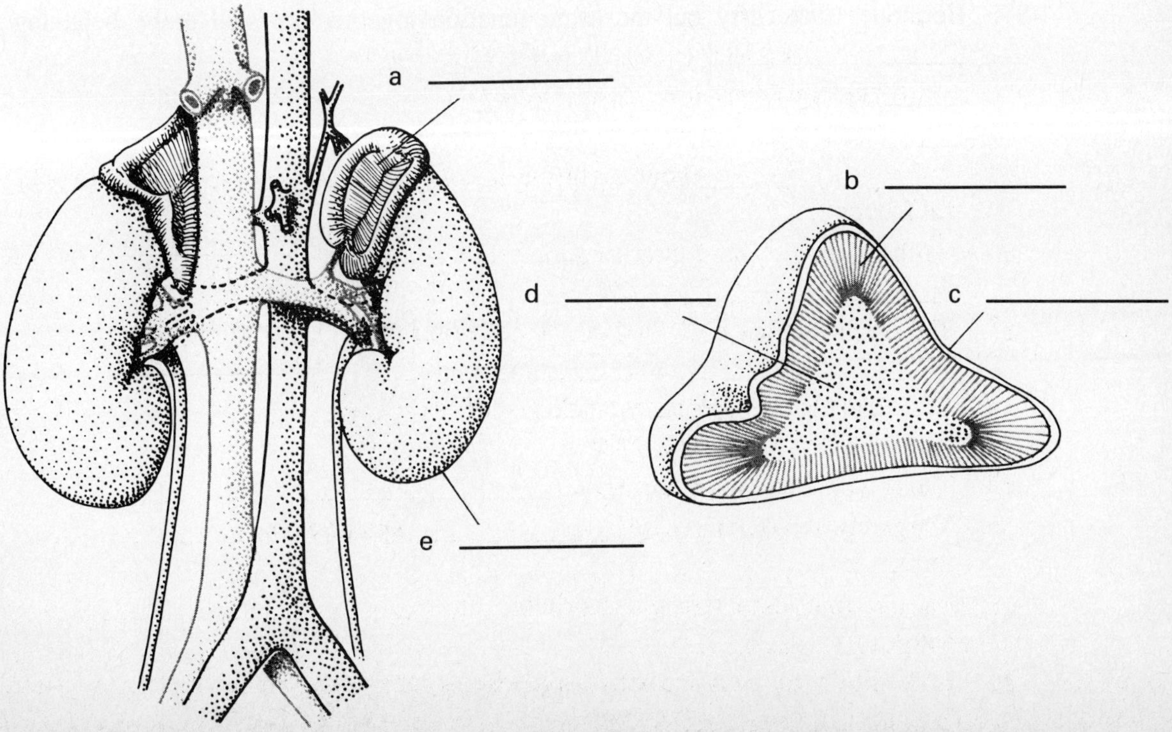

a _____

b _____

d _____

c _____

e _____

Figure 18.4

69. The alpha cells of the pancreas produce _____ and the beta cells produce _____.

70. _____ increases the transport of sugar across cell membranes thereby lowering sugar in the blood.

71. Insulin stimulates _____ and _____ formation.

72. Insulin dependent diabetes melitus is type _____.

73. Glucagon causes a (an) _____ in blood sugar.

74. Both glucagon and insulin are regulated primarily by the level of _____ in the blood.

75. Label figure 18.5, pancreatic tissue, using the terms listed below. Color each structure. Refer to figure 18.14 in your textbook for assistance.

blood vessel   alpha cells   beta cells   acinar cells

76. The principal male androgen is _____.

77. Testosterone is responsible for sperm maturation, maintenance of secretory glands, and _____ sex characteristics.

78. _____ suppresses FSH secretion.

79. Follicle cells surrounding the egg produce _____.

80. The _____ produces both estrogen and progesterone.

81. Estrogen supports _____ of the egg and growth of the uterine lining.

82. _____ prepares the uterus for the arrival of the embryo.

83. The sex hormones are regulated by the regulatory hormones of the _____.

84. The hormone of the pineal gland is _____.

85. Melatonin blocks the release of _____.

86. Melatonin reduces the rate of maturation of the sperm and egg by reducing the levels of _____.

87. Hormones that carry out the same function together are said to be behaving _____.

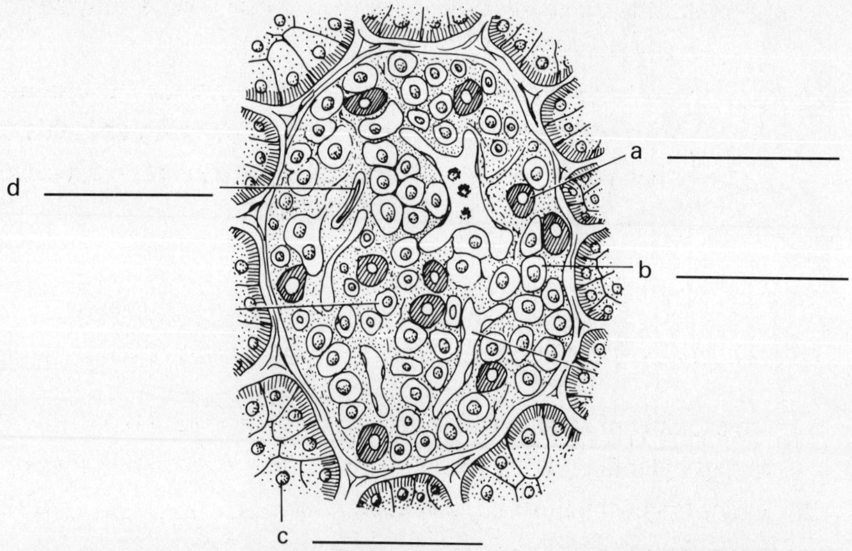

Figure 18.5

88. List the five hormones that are necessary for growth.

_____

_____

_____

_____

_____

89. Any threat to homeostasis represents _____.

90. The _____ represents the endocrinological and physiological responses to stress.

91. The dominant hormones involved in the alarm phase of the GAS are the _____.

92. During the resistance phase of the GAS the _____ hormones predominate.

93. The three major integrated results of the resistance phase are

_____

_____

_____

94. The malfunctioning of nerve and muscle cells marks the begining of the _____ phase of the GAS.

95. Circulating hormones have a substantial effect on _____.

96. Match the clinical condition listed below with the appropriate descriptive statement which follows. Refer to table 18.6 in your textbook for assistance.

a. dwarfism   b. acromegaly   c. diabetes insipidus   d. cretinism   e. Grave's disease   f. diabetes melitus type I   g. Addison's disease   h. Cushings syndrome   i. gynecomastia

_____ abnormal breast enlargement in males

_____ disease that results from an overproduction of thyroxine

_____ underproduction of insulin leads to this disease

_____ overproduction of the adrenal cortex hormones causes this

_____ occurs due to an underproduction of thyroxine during the growth period

_____ failure in ADH production results in this

_____ insufficient quantities of growth hormone cause this condition

_____ overproduction of growth hormone in adults results in this deformity

_____ underproduction of the hormones of the adrenal cortex cause this condition

SELF TEST

Circle the correct answer to each of the following questions.

1. An example of an amino acid derived hormone would be

a. prostaglandin.   b. thyroxine.   c. cortisol.   d. growth hormone   e. ADH.

2. Every target cell must have an appropriate _____ if it is to be affected by a given hormone.

a. enzyme   b. membrane   c. receptor   d. size   e. shape

3. The most important second messenger is

   a. adenyl cyclase. b. cyclic AMP. c. cyclic GTP. d. aldosterone. e. angiotensin II.

4. Steroid hormones exert their influence by

   a. activating enzymes. b. phosphorylating enzymes. c. activating genes. d. increasing cyclic AMP. e. none of the above.

5. Compared to the nervous system, the endocrine system as a control system

   a. is fast acting. b. affects fewer cells. c. is longer lasting. d. terminates faster. e. more than one of the above is correct.

6. The hormones of the anterior pituitary are regulated by the

   a. pineal gland. b. thyroid gland. c. infundibulum. d. hypothalamus. e. medulla.

7. Releasing and inhibiting factors reach the adenohypophysis by

   a. axonal transport. b. diffusion. c. the hypophyseal portal system. d. action potentials. e. neurotransmitters.

8. ADH is produced by the

   a. neurohypophysis. b. adenohypophysis. c. supraoptic nucleus.
   d. paraventricular nucleus. e. none of the above.

9. The target tissue of ACTH is the

   a. thyroid gland. b. adrenal medulla. c. adrenal cortex. d. gonads. e. pineal gland.

10. The hormone which causes an overall increase in metabolism is

    a. thyroxine. b. epinephrine. c. aldosterone. d. ADH. e. none of the above.

11. _____ causes an increase in serum calcium.

    a. ADH. b. PTH. c. Calcitonin d. ACTH. e. ANF.

12. Sodium retention is promoted by

    a. PTH. b. ACTH. c. ANF. d. aldosterone. e. somatostatin.

13. Red blood cell production is stimulated by the hormone released by the

    a. heart. b. pineal gland. c. kidney. d. adrenal gland. e. gonad.

14. The pineal gland produces

    a. thymosin. b. melatonin. c. erythropoietin. d. renin. e. ANF.

15. Underproduction of the adrenal cortex results in

    a. Cushing's disease. b. acromegaly. c. cretinism. d. Grave's disease
    e. Addison's disease.

CHAPTER 19
The Cardiovascular System: The Blood

OVERVIEW

We now begin the consideration of a new area, the circulating fluids of the body, and the organs associated with them. This chapter introduces you to the principal circulating fluid, the blood. It is the blood that carries oxygen and nutrients to the cells and removes the metabolic wastes from the cells. Consequently it is the key tissue for the maintenance of cellular homeostasis. In this role it serves as a connecting link for several other organ systems. It receives nutrients from the digestive system which is transported to the cells. Likewise, it receives oxygen from the respiratory system which is also transported to the cells. It removes and transports metabolic wastes to the excretory organs, the lungs for carbon dioxide, and the kidney for most other metabolic waste products. It assists in regulating cellular activity by transporting hormones to the various target cells.

In addition to these transport functions, the blood also plays a major role in acid-base balance, defense against microbial invasion, and regulation of body temperature.

This chapter begins with a listing of the blood's many functions and then follows this up with a detailed look at the composition of this complex tissue. Both the cellular and non-cellular components are examined. From there you will move to the clotting system which prevents excessive blood loss in cases of cardiovascular damage, and finally, the origin and generation of blood cells is considered.

Keep in mind as you read this chapter and the ones that follow that the blood is the functional component in the cardiovascular system. The heart and the blood vessels simply represent plumbing for distributing the blood. People do not die from hearts that have stopped, they die from lack of circulating blood!

CHAPTER OUTLINE

    A. Introduction
    B. The composition of blood
        1. The plasma
           a. The plasma proteins
        2. Cellular components
           a. Red blood cells
               (1) The hemoglobin molecule and gas transport
               (2) Blood types
               (3) Testing for compatibility
           b. White blood cells
               (1) Granular leukocytes
               (2) Agranular leukocytes
           c. Platelets
    C. The clotting system
    D. Hemopoiesis
        1. Erythropoiesis
        2. White blood cell formation
    E. Clinical patterns
        1. Disorders affecting the erythrocyte population
           a. Disorders involving abnormal numbers of erythrocytes
           b. Disorders involving abnormal erythrocyte structure
        2. Problems affecting white blood cell populations
        3. Problems with the clotting system

LEARNING ACTIVITIES

    Complete the following items by supplying the appropriate word or phrase.

    1. List four categories of substances that are transported by the blood.

        _____Oxygen_____

        _____

        _____

        _____

    2. List four other functions of the blood not involving transport.

        _____defense_____

        _____

        _____

        _____

    3. The ground substance of blood is the ____plasma____.

    4. The second component of whole blood are the ____formed____ elements.

    5. An adult man has _____5-6_____ liters of blood and an adult woman has ____4-5____ liters.

    6. A ____hypervolemic____ person would be one with a greater than normal blood volume.

7. Plasma contributes approximately _____55_____ percent of the volume of whole blood.

8. ____water____ makes up 92 percent of the plasma volume.

9. In terms of its composition, plasma tends to resemble interstitial fluid except that the concentrations of _____ and _____ are both higher in the plasma.

10. The three primary classes of plasma proteins are the _albumen_, the _____, and the _____.

11. The major contributors to the osmotic pressure of plasma are the _____.

12. The globulins are made up of transport proteins and _____.

13. Transport proteins that bind metal ions are known as _____.

14. Because most proteins are to large to pass across cell membranes, transport proteins keep small molecules from being lost in the _____.

15. Globulins which are involved in lipid transport are called _____.

16. The protein _____ functions in the clotting reaction.

17. The three types of formed elements are the _____, the _____, and the _____.

18. The _____ indicates the percentage of whole blood occupied by cellular elements.

19. The hematocrit range for an adult man is _____ and for an adult woman it is _____.

20. The average number of red cells per microliter of blood is _____ million.

21. In shape, the RBC has a _____ outer rim and a _thick_ center.

22. The strange shape of the RBC gives it a relatively large _surface_ area for rapid diffusion.

23. The shape of the RBC permits them to form stacks like dinner plates, which are termed _rouleaux_.

24. RBCs lack mitochondria, ribosomes, and _____.

25. The typical life span of an RBC is _____120_____ days.

26. Approximately _____ new RBCs enter the circulation each second.

27. _____ constitutes 95 percent of the protein found in a RBC.

28. Each hemoglobin molecule is made up of four subunits with each subunit consisting of a _____ and a _____.

29. It is the heme group that contains _____.

30. When environmental oxygen levels are high hemoglobin will saturate with oxygen and release it when environmental __O2__ levels are low.

31. When hemoglobin is broken down by phagocytic cells, the heme is converted into _bilirubin_.

32. Iron released into the blood stream by phagocytic cells binds to the plasma protein _____.

33. Excess iron is stored in the protein-iron complexes _ferritin_ and _hemosiderin_.

34. Complete the chart below. Refer to figure 19.5 in your textbook for assistance.

BLOOD TYPE	AGGLUTINOGEN	AGGLUTININ
A	A	anti-B
B	_____	_____
AB	A and B	_____
O	none	_____

35. For an individual with Rh negative blood to develop agglutinin against the Rh ag-glutinogen, they must be _____ to Rh positive blood.

36. A drop of blood is mixed with a drop of anti-A and cells can be seen clumping. A second drop of the same blood is mixed with anti-B and again cells clump. The blood type is _____.

37. Another two drops of blood from a different person are mixed with anti-A and anti-B as in number 36, but this time neither drop shows any cell clumping. The blood would be type _____.

38. A drop of blood is mixed with anti-D and shows no clumping. The person is Rh_____.

39. A universal donor would have _____ blood.

40. Erythroblastosis fetalis can only occur when an Rh _____ woman has a child by a Rh_____ man.

41. There are two major classes of white blood cells, the _____ and the _____.

42. All leukocytes are components of the _____ system.

43. Most of the white cells of the body are found in the _____ tissues, and the circulating leukocytes represent a small percentage of the population.

44. A typical microliter of blood would contain _____ thousand leukocytes.

45. Match the white blood cell listed below with the appropriate descriptive statement which follows.

    a. neutrophil b. eosinophil c. basophil d. monocyte e. lymphocyte

    _____ upon entering the peripheral tissues they are known as free macrophages

    _____ because of their strange shaped nuclei, these cells are frequently termed polymorphonuclear or PMNs

    _____ the most abundant white cell in the blood

    _____ the second most abundant white cell in the blood

    _____ representing about 2 - 4 percent of the total white cells, these cells are attracted to foreign compounds that have reacted with circulating antibodies

    _____ release histamine at sites of inflammation

    _____ numbers often increase dramatically during an allergic reaction

    _____ usually the first white cells to arrive at the site of injury

    _____ these cells produce antibodies

46. The precursor cell for the platelets is the _____.

47. Platelets are not true cells, but rather packets of _____.

48. Color figure 19.1, the blood cells. Use the colors that the cells appear when stained by standard methods. Refer to figures 19.2 and 19.8 in your textbook for assistance.

49. The processes involved in the prevention of blood loss is termed _____.

GRANULOCYTES   AGRANULOCYTES

Eosinophil   Basophil   Neutrophil   Lymphocyte   Monocyte

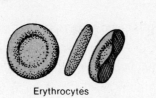

Erythrocytes   Platelets

Figure 19.1

50. The first phase of hemostasis occurs immediately after a blood vessel has been cut in is the ___Vascular___ spasm phase.

51. The second or ___platelet___ phase involves formation of a platelet plug that may seal the break.

52. The third phase is the ___coagulation___ phase where blood begins to clot.

53. The coagulation of blood involves the formation of a _____ mesh-work from the protein fibrinogen which then traps blood cells and forms a clot.

54. The clotting factors necessary for coagulation to occur include calcium and thirteen different proteins termed _____.

55. There are two different triggering mechanisms for initiating the clotting mechanism. One is the intrinsic pathway and the other is the _____ pathway.

56. Complete the following description of events that occur in the extrinsic pathway of clotting.

    Damaged cells release tissue factor which combines with __a__ to form tissue throm-boplastin. Tissue thromboplastin along with other factors activate the proenzyme __b__ converting it into thrombin. Thrombin then converts __c__ into fibrin.

    a. _____

    b. _____

    c. _____

57. Contraction of the platelets in the clot cause _____ which pulls the torn edges closer together.

58. Following the healing process, the clot is dissolved by the enzyme _____.

59. _____ is a drifting intravascular clot.

60. Three anticoagulants are _____, _____, and _____.

61. The enzymes _____ or _____ may be used to dissolve an embolus.

62. The process of blood cell formation is termed ___erythropoiesis___.

63. The primary site of erythropoiesis in the adult is the _____ tissue found in the bone marrow.

64. The signal for the release of erythropoietin is low _____ levels in the peripheral tissues, especially the kidney.

65. The stem cell from which all blood cells are formed is the _____.

66. The hemocytoblast gives rise to the _____ which is the first step in erythrocyte formation.

67. The _____ is the final stage of erythrocyte formation, just preceding the mature RBC.

68. All of the white cells belonging to the _____ class complete their development in the myeloid tissue.

69. The _____ are produced in peripheral tissues as well as in the bone marrow.

70. Match the clinical disorder listed below with the appropriate descriptive statement which follows.

a. polycythemia   b. hemorrhagic anemia   c. aplastic anemia   d. iron deficiency anemia   e. pernicious anemia   f. thalassemia   g. leukopenia   h. leukemia   i. thrombocytopenia

_____ an inadequate number of white cells

_____ this condition results from an abnormally low platelet count

_____ caused by an abnormally high hematocrit

_____ results from a dietary deficiency

_____ type of anemia that results from a shutdown of the bone marrow

_____ genetic defect that affects the shape of the hemoglobin molecule

_____ can be due to cancer of the myeloid or lymphoid tissues

_____ anemia due to blood loss

_____ inability to absorb vitamin B12 causes this type of anemia

SELF TEST

1. Blood functions in

   a. transport of gases.   b. transport of hormones.   c. transport of nutrients.   d. pH balance.   e. More than one of the above is correct.

2. After water, the most abundant constituent of plasma is the

   a. electrolytes.   b. proteins.   c. nutrients.   d. dissolved gases.   e. metabolic waste products.

3. A major function of albumin is the

   a. defense against microorganisms.   b. maintenance of blood osmotic pressure.
   c. prevention of blood loss.   d. generation of blood cells.   e. none of the above.

4. The clotting protein is

   a. albumin.   b. metalloprotein.   c. ferritin.   d. fibrinogen.   e. plasmogen.

5. The total percentage volume of cells in the blood is the

   a. red cell count.   b. white cell count.   c. differential count.   d. hematocrit.
   e. more than one of the above is correct.

6. There are _____ subunits in each hemoglobin molecule.

   a. 2   b. 3   c. 4   d. 5   e. 6

7. Heme is converted into _____ during the breakdown of hemoglobin.
   a. transferrin   b. ferritin   c. plasmogen   d. bilirubin   e. porphyrin

8. A person with type A blood would have.

   a. the B agglutinogen   b. the D agglutinogen   c. the A and B agglutinogen.   d. the A agglutinin.   e. none of the above

9. Under which of the following combinations of man and woman would a Rh compatibility problem be possible?

   a. negative man, positive woman   b. negative man, negative woman   c. positive man, positive woman   d. positive man, negative woman   e. more than one of the above is correct.

10. The most abundant white cell is the

    a. neutrophil.   b. basophil.   c. eosinophil.   d. lymphocyte   e. monocyte.

11. The white cell which produces antibody is the

    a. neutrophil.   b. basophil.   c. eosinophil.   d. lymphocyte.   e. monocyte.

12. Platelets are derived from

    a. reticulocytes.   b. promonocytes.   c. megakaryocytes.   d. normoblasts.
    e. myelocytes.

13. The first phase of hemostasis is

    a. coagulation.   b. vascular spasm.   c. clot retraction.   d. platelet phase.   e. intrinsic.

14. Prothrombin is converted into thrombin by

    a. fibrinogen.   b. fibrin.   c. factor XII.   d. thromboplastin.   e. plasmogen.

15. The stem cell for all of the blood cells is the

    a. normoblast.   b. megakaryocyte.   c. hemocytoblast.   d. reticulocyte.
    e. erythropoietin.

CHAPTER 20
The Cardiovascular System: The Heart

OVERVIEW

It was pointed out in the last chapter that the functions that we normally associate with the cardiovascular system are really the functions of blood. The remainder of the system, the heart and blood vessels, simply exist to keep the blood flowing. Of course if the heart or vessels fail then the blood no longer flows and death follows in a few minutes. In this chapter we will consider the motive force that keeps the blood flowing, the heart. As organs go, the heart is relatively simply, being nothing more than a pump. In spite of its simplicity, if the pump fails then death follows quickly. Heart disease and failure is the leading cause of death in the United States each year. Consequently a tremendous amount of research has gone into preventing heart disease and to repairing or replacing defective hearts.

This chapter begins with an overview of the circulation. This is followed by a detailed look at the gross anatomy of the heart. From there you will proceed to a brief consideration of cardiac muscle histology and then the remainder of the chapter is devoted to the physiology of the heart. Within the context of physiology you will examine the electrochemical aspects of cardiac muscle, the cycle of cardiac activity, and cardiodynamics, or how the amount of blood pumped by the heart is controlled.

As you proceed through this chapter you should always keep in mind that the function of the heart is to pump as much blood as is needed by the tissues of the body.

CHAPTER OUTLINE

    A. Introduction
    B. Gross and functional anatomy
        1. Location and external appearance

    2. Sectional anatomy and organization
    3. Coronary circulation
    4. Cardiac histology
  C. Cardiac physiology
    1. Action potentials and cardiocyte membrane
    2. The coordination of cardiac contractions
    3. The extracellular fluid and cardiac function
    4. The heart as a pump
      a. The cardiac cycle
      b. Valve structure and function
      c. Heart sounds
      d. The electrocardiogram (ECG)
    5. Cardiodynamics
      a. The control of heart rate
      b. The regulation of stroke volume
        (1) The EDV and intrinsic regulation of stroke volume
        (2) The ESV and autonomic regulation of stroke volume
        (3) The coordination of autonomic activity
  D. Clinical patterns
    1. Developmental defects
    2. Infection
    3. Cardiomyopathies

LEARNING ACTIVITIES

Complete each of the following items by supplying the missing word or phrase.

1. The major vessels of the circulatory system include the ___Veins___, ___arteries___, and ___capillaries___.

2. The circulatory circuit that carries blood to and from the lungs is the ___Pul___ circuit, while the ___Systemic___ circuit conveys blood to the rest of the body.

3. The right side of the heart receives blood from the ___Systemic___ circuit and the left side receives it from the ___pulmonary___ circuit.

4. Each side of the heart consists of a thin walled ___atrium___ and a thick walled ___ventricle___.

5. The heart is found in a connective tissue surrounded space in the thoracic cavity termed the ___mediastinum___.

6. The heart lies within the ___pericardial___ cavity.

7. The pericardium consists of two membranes, an inner, ___visceral___ pericardium (epicardium), and an outer ___pericardium___ pericardium.

8. Both pericardial membranes are ___serous___, and secrete fluid into the pericardial cavity, which is the gap between them.

9. The pericardial fluid acts as a ___lubricant___ between the two membranes.

10. Label figure 20.1, the pericardial membranes, using the terms listed below. Color each structure. Refer to figure 20.2 in your textbook for assistance.

    parietal pericardium   visceral pericardium   pericardial space
    fibrous layer   serous layer   myocardium

11. The atria are separated from the ventricles by the ___coronary___ sulcus.

Label: parietal pericardium   visceral pericardium   pericardial space   fibrous layer   serous layer
myocardium

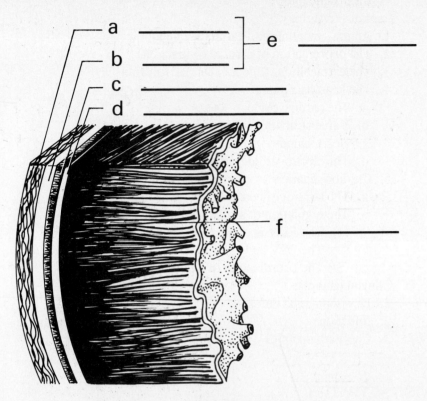

a _____   ⎤
                   ⎬ e _____
b _____   ⎦

c _____

d _____

f _____

Figure 20.1

12. The ventricles are separated from one another by the ~~Interventricular Septum~~ sulcus.

13. Internally, the atria are separated by the Interatrial septum and the ventricles are separated by the Interventricular septum.

14. Match the terms listed below with the appropriate descriptive statements which follow.

a. vena cavas b. pectinate muscle   c. tricuspid valve   d. chordae tendinae e. papillary muscle   f. trabeculae carnae   g. pulmonary trunk   h. pulmonary veins i. bicuspid valve   j. pulmonary semilunar valve   k. aortic semilunar valve   l. aorta

___L___ receives the blood ejected by the left ventricle

___C___ valve that separates the right atrium from the right ventricle

___D___ fibrous braces of the AV valves

___J___ valve that separates the right ventricle from the pulmonary artery

___E___ muscles on the walls of the ventricles to which the chordae tendinae are attached

___H___ veins that convey blood to the left atrium

___I___ valve that separates the aorta from the left ventricle

___G___ receives the blood ejected from the right ventricle

___B___ muscular ridges found on the inner walls of the atria

___F___ deep grooves and folds found on the internal wall of the ventricles

15. Label figure 20.2, the internal structure of the heart, using the terms listed below. Color the pathways of deoxygenated blood blue and the pathways of oxygenated blood red. Refer to figure 20.4 in your textbook for assistance.

superior vena cava   right pulmonary artery   right pulmonary veins
pulmonary semilunar valve   tricuspid valve   chordae tendinae
inferior vena cava   papillary muscle   left ventricle
right atrium   aortic semilunar valve   left pulmonary veins
left atrium   left pulmonary artery   aorta
right ventricle   bicuspid valve

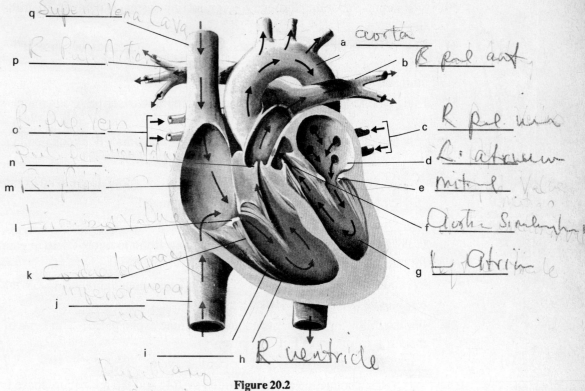

q _Super Vena Cava_

p _R Pul Arto_

o _R Pul Vein_

n _____

m _R pul_

l _Tricuspid value_

k _Chordae Tendinae_

j _Inferior Vena Cava_

i _____   h _R ventricle_

a _aorta_

b _R Pul aat_

c _R Pul Veins_

d _L atrium_

e _mitral value_

f _Aortic Semilunar value_

g _Lt Atrium_

papillary muscle

Figure 20.2

16. The heart is supplied with blood by the _Coronary_ circulation.
17. The right coronary artery gives rise to the _marginal_ branch and the posterior _interventricular_ branch.
18. The left coronary artery gives rise to the _Circumflex_ branch and the anterior _interventricular_.
19. The condition of coronary _Ischemia_ results from partial or complete blockage of a coronary artery.
20. Transitory chest pain due coronary ischemia is known as _Angina Pectoris_.
21. Another name for a heart attack is _Myocardial_ infarction.
22. Myocardial infarction is due to a blockage of the _Coronary_ circulation and the subsequent death of heart tissue.
23. The three layers of the heart are the _epicardium_, _endocardium_, and _myocardium_.
24. Cardiocytes make contact with each other at specialized sites known as _Intercalated_ discs.
25. The refractory period of cardiocyte membranes is about _300_ milliseconds as compared to about 10 milliseconds for skeletal muscle.
26. The long refractory period of the cardiocyte membrane means that it is impossible to drive the heart into a _tetanus_ contraction.

27. The first nodal cells to depolarize are the _Pacemaker_ cells and will determine the rate at which the heart contracts.

28. Complete the sequence of conduction through the heart.
    SA node - atrial conducting fibers - _AV node_ - Bundle of His - _Lt R Bun Brch_ - Purkinje cells

29. The bundle of His is located inside of the _Interventricular_ septum.

30. The impulse traveling through the conducting system is delayed 50 milliseconds at the _AV node_.

31. The delay of the signal by the AV node is important because it permits the _atria_ to complete contraction before the _ventricles_ are stimulated.

32. Match the term listed below with the appropriate descriptive statement which follows.
    a. hypercalcemia   b. hypocalcemia   c. hyperkalemia   d. hypokalemia
    _D_ causes heartbeats to become very weak
    _C_ will result in hyperpolarization and cause a slow down in heart rate
    _A_ heartbeat becomes powerful and prolonged
    _B_ results in a weak and irregular heartbeat

33. Label figure 20.3 using the terms listed below. Refer to figure 20.10 in your textbook for assistance.
    SA node   Bundle of His   Right-bundle branch   AV node
    left-bundle branch   Purkinje fibers

34. For the following statements about the cardiac cycle, place a T in front of the statement if it is true and a F if it is false.
    _T_ There are two phases to the cardiac cycle.
    _T_ The contraction phase is termed systole and the relaxation phase is termed diastole.
    _F_ Systole is the longer of the phases in the cardiac cycle.
    _T_ The ventricles fill to 70% capacity by gravity.
    _F_ Ventricular systole begins before atrial systole.
    _T_ The AV valves open when atrial pressure exceeds ventricular pressure.
    _F_ The aortic semilunar valve closes when the pressure in the left ventricle exceeds the aortic pressure.
    _T_ It is the valves of the heart that control the direction of flow.
    _T_ The AV valves close when the ventricular pressure exceeds the atrial pressure.

35. The cusps of the AV valves are prevented from swinging back into the atria by tension on the _Chordae Tendinae_ by the papillary muscles.

36. The closing of the _AV_ valves is responsible for the first heart sound.

37. The second heart sound occurs at the beginning of _ventricular_ diastole.

38. The mitral valve has _two_ cusps.

39. Label figure 20.4, valves of the heart, using the terms listed below. Color each valve. Refer to figure 20.13 in your textbook for assistance.
    pulmonary semilunar valve   aortic semilunar valve   pulmonary trunk
    aorta   bicuspid valve   tricuspid valve

40. A recording of the electrical activities of the heart is the _ECG_.

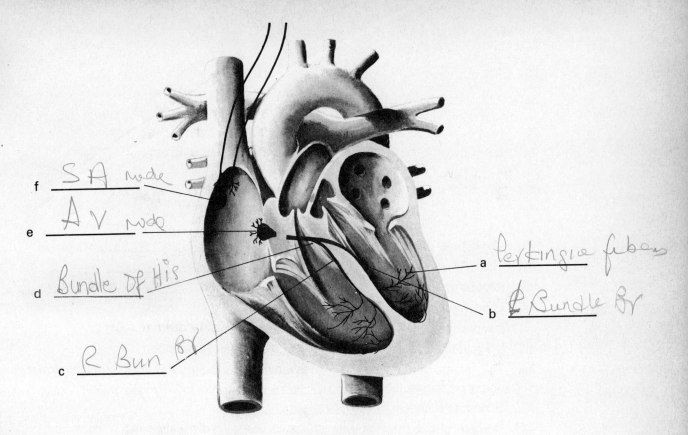

f ___ SA node ___
e ___ AV node ___
d ___ Bundle Of His ___
c ___ R Bun Br ___

a ___ Perkinjie fibers ___
b ___ L Bundle Br ___

Figure 20.3

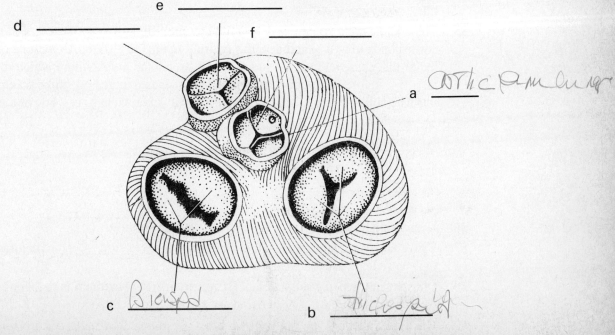

d ___
e ___
f ___

a ___ aortic semilunar ___

c ___ Bicuspid ___
b ___ Tricuspid ___

Figure 20.4

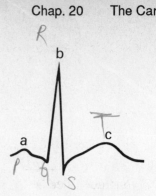

Figure 20.5

41. The depolarization of the atria is represented by the _____P_____ wave.
42. The QRS wave represents the depolarization of the _ventricle_.
43. Repolarization of the ventricles is represented by the _____T_____ wave.
44. Label figure 20.5, an ECG segment, using the following terms.

    P wave   QRS wave   T wave

45. CO (cardiac output) = _Stroke Vol_ X HR (heart rate)
46. The volume of blood between the resting cardiac output and the maximum cardiac output represents the _Cardiac Resorve_
47. Heart rate is under _Autonomic_ control.
48. Norepinephrine from the sympathetic nerves _Increases_ heart rate.
49. Acetylcholine from branches of the vagus nerve _Decreases_ heart rate.
50. Cutting the vagus nerve will _Increase_ heart rate.
51. Blocking the sympathetic nerves to the heart will _Slow_ heart rate.
52. Stretching the walls of the right atrium by increasing blood volume will cause an increase in heart rate. This is known as the _atrial_ reflex.
53. SV (stroke volume) = EDV - _ESV_
54. At the end of diastole a heart contains 130 ml of blood. Following systole there was 60 ml left in the ventricle. The stroke volume of the heart was _70_ ml.
55. The end diastolic volume is determined by the filling time and __a__. Filling time corresponds to __b__ and depends entirely upon __c__ rate. Alterations of either factor that cause an increase in EDV will stretch the myocardium which results in an __d__ in force of contraction, which in turn causes increased stroke volume. This intrinsic ability of the heart to adjust its output is known as __e__ law of the heart.

    a. _Venous Return_
    b. _diastole_
    c. _Heart Rate_
    d. _Increase_
    e. _Starlings_

56. Decreasing ESV will _Increase_ stroke volume.
57. ESV changes are largely a result in changes in the _force_ of contraction of the ventricles.
58. Changes in force of contraction of the ventricular myocardium is regulated mainly by the _Autonomic_ division and the adrenal medulla.
59. Heart rates above _180_ beats per minute leave such a short filling time that EDV drops and with it drops the stroke volume.

60. Autonomic control of cardiac output comes from _Cardiac_ centers in the medulla.

61. The _cadioaccellato_ center activates sympathetic nerves to the heart.

62. The _Cardioinhibitory_ center governs the activity of parasympathetic neurons to the heart.

63. The cardiac control centers receive information from the higher centers, chemoreceptors, and _Baroreceptors_.

64. Match the clinical condition listed below with the appropriate descriptive term that follows.

a. septal defect   b. rheumatic fever   c. endocarditis   d. cardiac tamponade   e. cardiac myopathy

___b___ usually caused by a streptococcal infection

___a___ permit blood to pass directly between the atria

___c___ infections which affect primarily the valves and chordae tendinae cause this condition

___d___ frequently occurs with pericarditis

___e___ cardiocytes die and are replaced by fibrous scar tissue

SELF TEST

Circle the correct answer to each of the following questions.

1. The heart is located in the
a. pleural cavities.   b. dorsal cavity.   c. diaphragm.   d. mediastinum.   e. None of the above.

2. Another name for the visceral pericardium is the
a. endocardium.   b. myocardium.   c. epicardium.   d. mediastinum   e. pericardial cavity.

3. Externally, the atria are separated from the ventricles by the
a. coronary sulcus.   b. intervetricular sulcus.   c. interatrial groove.   d. interventricular septum.   e. interatrial septum.

4. The vessels that open into the right atrium are the
a. pulmonary arteries.   b. pulmonary veins.   c. vena cavas.   d. coronary arteries.   e. coronary veins.

5. The muscular ridges of the anterior atrial wall are the
a. papillary muscles.   b. chordae tendinae   c. pectinate muscles   d. AV valves.   e. semilunar valve.

6. Backflow from the pulmonary artery is prevented by the
a. aortic semilunar valve.   b. tricuspid valve.   c. bicuspid valve.   d. pulmonary semilunar valve.   e. mitral valve.

7. The circumflex artery is derived from the
a. anterior interventricular artery.   b. left coronary artery.   c. right coronary artery.   d. posterior interventricular artery.   e. coronary sinus.

8. The reason that the heart cannot be driven into tetanus as can skeletal muscle is
    a. the magnitude of the action potential is greater in cardiac muscle.   b. the magnitude of the action potential is greater in skeletal muscle.   c. the refractory period is much longer in cardiac muscle.   d. the refractory period is much longer in skeletal muscle.   e. that there are not sufficient numbers of action potentials per minute to the heart.

9. The heartbeat begins at the
    a. AV node.   b. bundle of His.   c. SA node.   d. left bundle branch.   e. right bundle branch.

10. A very powerful and prolonged heart contraction might be due to
    a. elevated potassium levels.   b. low potassium levels.   c. elevated calcium levels.   d. low calcium levels.   e. low sodium levels.

11. The AV valves open when
    a. atrial pressure exceeds ventricular pressure.   b. ventricular pressure exceeds atrial pressure.   c. aortic pressure exceeds atrial pressure.   d. aortic pressure exceeds ventricular pressure.   e. pulmonary pressure exceeds aortic pressure.

12. The total volume of blood pumped per minute per ventricle is the
    a. heart rate.   b. stroke volume.   c. cardiac output.   d. ESV.   e. EDV.

13. Heart rate is controlled primarily by the
    a. atrial reflex.   b. Bainbridge reflex.   c. Starling's law.   d. autonomic nervous system   e. sinus arrhythmia

14. The EDV is 150 ml and the ESV is 75 ml. Heart rate is 100 BPM. Cardiac output will be
    a. 5000 ml.   b. 15,000 ml.   c. 7500 ml.   d. 10,000 ml.   e. none of the above.

15. Inflammation of the lining of the heart is known as
    a. pericarditis.   b. endocarditis.   c. cardiac tamponade.   d. cardiac arrest.   e. myocarditis.

CHAPTER 21
The Cardiovascular System: Vessels and Circulation

OVERVIEW

This chapter continues the theme of the cardiovascular system by examining the third major component of the system, the blood vessels. The heart is what moves the blood but it is the vessels that distributes the blood to all of the regions of the body. It is important to understand that the vessels are not simply passive "plumbing" but are dynamic living organs capable of adjusting the flow of blood as required by the various tissues of the body.

The chapter begins with a consideration of the structure of the various vessels and then moves to the gross anatomy of the circulatory system, the various vessels and the regions which they both supply and drain. Having established the major anatomical features of the system, the chapter then proceeds to consider the physiology of blood flow. This topic is sometimes referred to as hemodynamics and takes in all of the factors which contribute to flow. Flow is equal to the fluid pressure divided by the resistance offered to flow. The first aspect considered in this section is blood pressure, followed by peripheral resistance. As flow in the circulatory system is equal to the cardiac output, the relationship between cardiac output, blood pressure, and vessel resistance is explored. Once the factors affecting flow have been established, an examination of the regulation of these factors is undertaken. Finally patterns of cardiovascular response are considered with special emphasis on the effects of exercise and hemorrhaging or blood loss.

CHAPTER OUTLINE

A. Introduction
B. The functional anatomy of the circulatory system

1. Histological organization
   a. Arteries
   b. Capillaries
      (1) Capillary beds
   c. Veins
2. The distribution of blood

C. The circulatory system
   1. The pulmonary circulation
   2. The systemic circulation
      a. The arterial system
         (1) Arteries originating on the aortic arch
            (a) The subclavian artery
            (b) The carotid artery and the blood supply to the brain
            (c) The descending aorta
         (2) The venous system
            (a) The superior vena cava
            (b) The venous return from the limbs and chest
            (c) The inferior vena cava
            (d) The hepatic portal system
   3. The development of the circulatory system

D. Circulatory physiology
   1. Blood pressure
      a. Arterial pressures
      b. Venous pressures
   2. Peripheral resistance
   3. Circulatory dynamics
      a. Variations in cardiac output
      b. Variations in peripheral resistance
      c. Alterations in blood pressure

E. Homeostasis and cardiovascular function
   1. Neural mechanisms and the short-term regulation of blood pressure
      a. Barorecptor reflexes
      b. Chemoreceptor reflexes
      c. Autonomic activation and higher centers
   2. Hormones and cardiovascular regulation
   3. Local factors affecting blood volume and blood pressure
      a. The dynamic center

F. Patterns of cardiovascular response
   1. Exercise and the cardiovascular system
      a. Cardiovascular fitness
   2. The cardiovascular response to hemorrhaging
      a. The elevation of blood pressure
         (1) Mobilizing venous reserve
         (2) Hormonal responses
         (3) The central ischemic response
         (4) The restoration of blood volume

G. Clinical patterns
   1. Hypertension
   2. Hypotension

LEARNING ACTIVITIES

Complete each of the following items by supplying the appropriate word or phrase.

1. The three layers that make up the walls of the arteries and veins are the _T. media_, _T. External_, and _T. Interna_.

2. The tunica interna consists of an _endothelium_ lining resting on an elastic connective tissue membrane.

3. The smooth muscle of the blood vessel wall is concentrated in the _T. Media_.

4. Collagenous fibers dominate the _T. External_ layer.

5. Label figure 21.1, structure of arteries and veins, using the terms listed below. Color each layer. Refer to figure 21.1 in your textbook for assistance.

tunica externa	tunica media	elastic membrane
endothelium	tunica intima	endothelial cells.

6. The _elastic_ arteries have the largest diameters of the arteries.

7. The tunica media of elastic arteries contains little smooth muscle but a large number of _elastic_ fibers.

8. Blood is distributed to the peripheral tissues by the _____ arteries.

9. The _T. Media_ have a tunica media composed entirely of smooth muscle.

10. The smooth muscle found in the walls of arterioles and muscular arteries permits them to adjust their _Diameter_.

11. The _arterioles_ are the small blood vessels that supply the tissues of the thick walls of the large arteries and veins.

12. The wall of a capillary consists of an _endothelium_.

13. The average diameter of a capillary is _____ micrometers.

14. _fenestrated_ capillaries are found in the kidney and the choroid plexus.

15. Flattened irregular capillaries found in the liver, bone marrow, and the adrenal glands are known as _Sinusoids_.

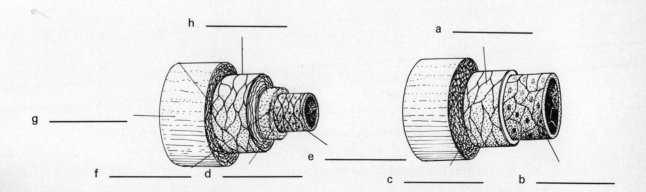

Figure 21.1

16. Arterioles supply _Capillary_ beds.

17. A band of smooth muscle at the entrance of each capillary is known as the _precapillary sphincter_ and permits regulation of blood flow into the capillary.

18. _____ is the alternative relaxing and constricting of the precapillary sphincters, and results in pulses of blood flow.

19. Multiple arterioles that supply a given capillary bed are termed _____, and insure a constant blood supply even if one becomes blocked.

20. _____ constitute a preferential shunt from an arteriole to a vennule.

21. The thickest layer of large veins is the tunica _____.

22. There are _____ in many veins that prevent the backflow of blood.

23. The bulk of the blood volume is contained on the _____ side of the circulation.

24. The venous system acts as a blood _____, and can supply blood rapidly to the arterial side in case of emergency.

25. _____ are varicose veins in the rectum.

26. Complete each of the following blood flow sequences by supplying the missing vessel. Each sequence traces a drop of blood through all of the major vessels through which it would pass on its way to a particular part of the body and back. The beginning point is the left ventricle and the end point is the right atrium.

    A. Foot LV - aorta - common iliac artery - (a)_____ _____ - (b)_____ _____ - popiteal artery - (c)_____ _____ - foot - great saphenous vein - (d)_____ _____ - external iliac vein - (e)_____ _____ - inferior vena cava - RA

    B. Small intestine LV - aorta - (a)_____ _____ - small intestine - superior mesenteric vein - (b)_____ _____ - liver - (c)_____ _____ - inferior vena cava - RA

    C. Right occipital lobe of the brain LV - aorta - (a)_____ _____ - right common carotid - (b)_____ _____ - posterior communicating artery - (c)_____ _____ - occipital lobe - sinus - internal jugular vein - (d)_____ _____ - superior vena cava - RA

    D. Left hand LV - aorta - (a)_____ _____ - axillary artery (b)_____ _____ - ulnar artery - hand - radial vein - (c)_____ _____ - (d)_____ _____ - subclavian vein - (e)_____ _____ - superior vena cava - RA

27. The brachiocephalic artery gives rise to the _____ artery and the _____ artery.

28. The circle of Willis receives blood from the internal carotid arteries and the _____ artery.

29. The basilar artery is formed by the anastomosis of the _____ arteries.

30. The two major veins which contribute to the hepatic portal vein are the splenic vein and the _____ vein which drains the small intestines.

31. The stomach receives blood via the _____ artery, which branches off the aorta.

32. The face, scalp, and dura mater receive blood via the _____ artery.

33. The external jugulars, internal jugulars, and subclavian veins unite to form the _____ trunk.

34. The posterior surface of the arm and hand are drained by the _____ vein.

35. The three major veins that empty into the femoral vein are the great saphenous, popiteal, and _____ vein.

36. The _____ vein runs parallel to the vertebral column on the right side of the midline.

37. The lateral surface of the arm is drained by the _____ vein.

38. The bulk of the blood leaving the brain drains into the dural _____.

39. Label figure 21.2, the major arteries, using the terms listed below. Refer to figure 21.8 in your textbook for assistance.

left common carotid	right common carotid	internal carotid	pulmonary
external carotid	brachiocephalic	arch of aorta	celiac
superior mesenteric	common iliac	internal iliac	external iliac
popiteal	posterior tibial	anterior tibial	femoral
ulnar	radial	renal	brachial
left subclavian	axillary		

40. Label figure 21.3, the major veins, using the terms listed below. Refer to figure 21.11 in your textbook for assistance.

external jugular	internal jugular	ulnar	brachial
superior vena cava	inferior vena cava	hepatic	renal
common iliac	external iliac	femoral	great saphenous
lesser saphenous	popiteal	posterior tibial	anterior tibial
internal iliac	right brachiocephalic	basilic	radial
axillary	subclavian		

41. Label figure 21.4, the hepatic portal system, using the terms listed below. Refer to figure 21.15 in your textbook for assistance.

inferior mesenteric vein	superior mesenteric vein	portal vein
splenic vein	gastroepiploic vein	right gastric vein
inferior vena cava	hepatic vein	pyloric vein

42. In the fetus, the pulmonary circulation is bypassed by means of the _____ and the _____.

43. The umbilical arteries are extensions of the _____ arteries in the mother.

44. Blood returns to the fetus from the placenta via the _____ vein.

45. Blue babies result when the foramen _____ remains open after birth.

46. Flow rate is directly proportional to _____ applied to the fluid.

47. Flow rate is inversely proportional to the _____ that opposes the fluid.

48. Resistance to flow results from friction between the molecules of the fluid, friction between the fluid and the walls of the pipe, and _____.

49. Blood moves and circulates because the heart applies _____.

50. The presure difference between pressure in the aorta and pressure in the right atrium represents _____ pressure.

51. Arterial pressure has two values, an upper value termed the __a__ pressure, and a lower pressure termed the __b__ pressure. The difference between these pressures is the __c__ pressure. The average arterial pressure is the __d__ arterial pressure. As the blood moves away from the heart the pulse pressure becomes less, and dis-

Label: left common carotid   right common carotid   internal carotid   pulmonary   external carotid
brachiocephalic   arch of aorta   celiac   superior mesenteric   common iliac
internal iliac   external iliac   popiteal   posterior tibial   anterior tibial   femoral
ulnar   radial   renal   brachial   left subclavian   axillary

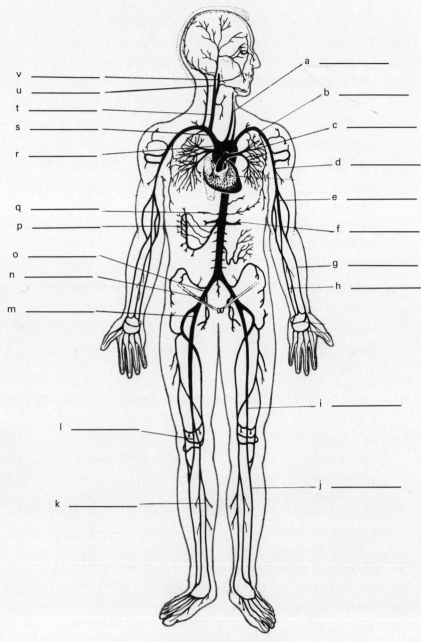

v _____
u _____
t _____
s _____
r _____

a _____
b _____
c _____
d _____
e _____
f _____
g _____
h _____

q _____
p _____
o _____
n _____
m _____

i _____

l _____

j _____

k _____

Figure 21.2

appears completely in the arterioles. This is because of the __e__ rebound of the
blood vessels which prevents the systolic pressure from rising to high, and the diastolic pressure from dropping to low.

a. _____

b. _____

c. _____

d. _____

e. _____

Label: external jugular   internal jugular   ulnar brachial   superior vena cava   inferior vena cava
        hepatic   renal   common iliac   external iliac   femoral   great saphenous
        lesser saphenous   popiteal   posterior tibial   anterior tibial   internal iliac
        right brachiocephalic   basilic   radial   axillary   subclavian

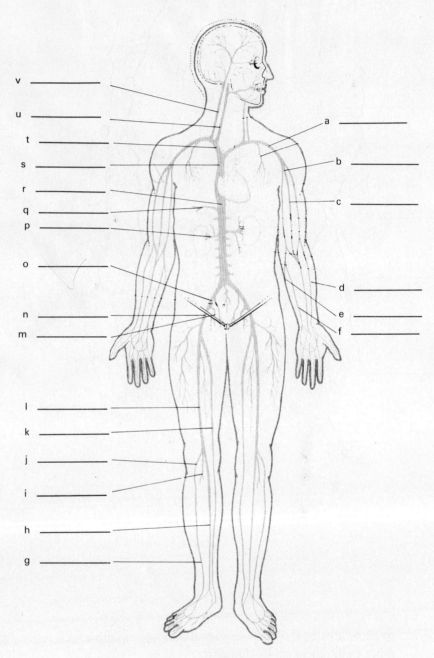

Figure 21.3

52. By the time blood reaches the capillaries, the mean arterial pressure is about
    _____ mm Hg.

53. Venous pressures are significant because they determine the _____ to
    the heart, and therefore cardiac output.

54. Venous blood has little __a__. In order to get blood back to the heart, against gravity,
    veins rely upon __b__ which prevent the backflow of blood, __c_ pumps generated

Label: inferior mesenteric vein   superior mesenteric vein   portal vein   splenic vein   gastroepiploic vein
          right gastric vein   inferior vena cava   hepatic vein   pyloric vein

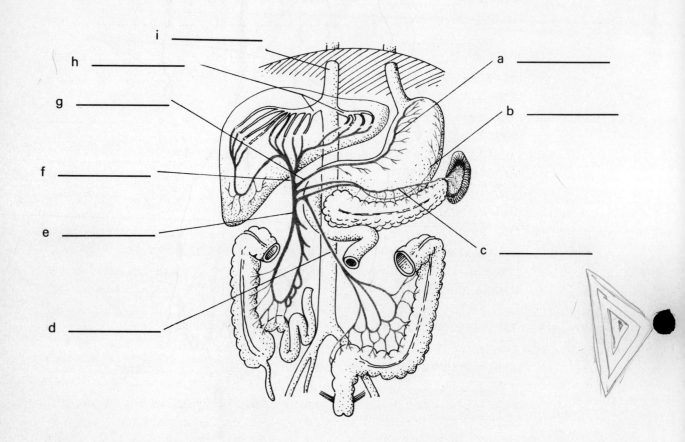

Figure 21.4

by skeletal muscle contractions, and the __d__ pump which is generated by pressure
drops in the chest cavity during inspiration.

a. _____

b. _____

c. _____

d. _____

55. The total peripheral resistance that the vascular system offers to blood flow is
     primarily composed of _____ resistance.

56. The peripheral resistance of the arterioles can be adjusted by relaxing the smooth
     muscle in the arteriole wall which causes _____ or contracting the
     muscle which causes _____.

57. The diameter of the arterioles is under control of the _____ center which is located in the medulla.

58. Stimulation of the vasomotor center causes _____.

59. For any fluid enclosed by a series of pipes, the flow rate is equal to the pressure applied divided by the resistance. When applied to the circulatory system, flow rate is equal to _____ which is the volume of blood flow per minute, pressure is equal to _____, and resistance is equal to _____.

60. Cardiac output is affected by _____, _____, and _____ factors.

61. The major local factor which affects cardiac output is _____ return.

62. Increasing cardiac output while holding peripheral resistance constant will cause a (an) _____ in blood pressure.

63. Local dilation of precapillary sphincters will be caused by _____ oxygen, _____ carbon dioxide, and _____ pH.

64. Local dilation of precapillary sphincters on a large scale will _____ peripheral resistance.

65. Inhibition of the vasomotor center will cause _____ and decrease peripheral resistance.

66. The hormones epinephrine, norepinephrine, ADH, and angiotensin cause _____ and increase peripheral resistance.

67. Increasing the concentration of proteins in the plasma or the hematocrit causes an increase in the _____ of the blood which in turn causes a (an) _____ in peripheral resistance.

68. If cardiac output remains constant, an increase in peripheral resistance will cause a (an) _____ in blood pressure.

69. Blood pressure varies directly with _____, _____, and _____.

70. In the neural control of blood pressure a number of different events occur which effect each other. For each event or activity listed below, indicate whether it will increase or decrease by placing an I or D in the space provided. The physiological condition that is to be responded to is an increase in blood pressure above the normal level.

    a. _____ baroreceptor activity
    b. _____ vasomotor activity
    c. _____ constriction of arterioles
    d. _____ peripheral resistance
    e. _____ cardioaccelerator center activity
    f. _____ cardioinhibitory center activity
    g. _____ cardiac output
    h. _____ blood pressure

71. Decreased oxygen levels in the blood which is detected by the _____ or _____ will lead to stimulation of the vasomotor and cardioacceleratory centers thereby elevating blood pressure and flow rates.

72. In the space provided before the statements listed below, write the name of the hormone that is responsible for each action.

    _____ stimulates the reabsorption of sodium and water, thereby increasing blood volume and blood pressure.

_____ stimulates the production of red blood cells.

_____ causes an immediate constriction of arterioles but has long term effects in that it promotes retention of water by the kidneys.

_____ increases water loss by the kidneys, blocks the release of ADH and aldosterone, and stimulates peripheral vasodilation

73. All exchange between the blood and the tissues occurs across the walls of the _____.

74. Fluid leaves the circulation when the hydrostatic (blood) pressure exceeds the _____ pressure of the plasma.

75. At the _____ hydrostatic and osmotic pressure are equal and no net movement of fluid occurs.

76. Past the dynamic center, osmotic pressure exceeds _____ pressure and fluid moves back into the capillary.

77. The fluid that is not pulled back into the circulation eventually returns via the _____ system.

78. A decrease in the plasma protein concentration would result in _____ volumes of tissue fluid.

79. _____ is where tissues have become swollen due to excess tissue fluid.

80. When exercise begins the precapillary sphincters of the capillaries in the muscles dilate and peripheral resistance drops. Blood pressure does not drop correspondingly because _____ increases as a result of increased venous return.

81. A regime of physical exercise usually results in a (an) _____ resting heart rate and a (an) _____ resting stroke volume.

82. Following hemorrhage, blood pressure is immediately maintained by increasing _____.

83. If the volume of blood lost during hemorrhage is not too severe, it can be compensated for by mobilizing the _____ reserve.

84. If the volume of blood lost is severe then the _____ response will shut down flow to all peripheral tissues except the brain.

85. For each statement below, place a T in front of it if it is true and a F if it is false.

_____ Hypertension is the correct name for high blood pressure.

_____ For a person to be classified as hypertensive, their pressure must exceed 160/100.

_____ Essential hypertension is due to a need for higher blood pressure.

_____ Secondary hypertension is caused by excessive production of hormones involved in water and sodium retention.

_____ Hypertension causes an increase in the size and mass of the heart.

_____ Baroreceptors in hypertensive individuals attempt to return the excessive pressure to normal.

_____ Diet and exercise can be used in some individuals to control hypertension.

_____ Initially, hypertension has no symptoms.

_____ Orthostatic hypotension can occur in normal individuals when they arise quickly from a prone position.

_____ Reduced cardiac output is the definition of shock.

_____ Cardiogenic shock is due to a loss of blood.

_____ Shutdown of the vasomotor center can result in neurogenic shock.

_____ Anaphylactic shock occurs due to peripheral vasodilation following an allergic response.

SELF TEST

Circle the correct answer to each of the following questions.

1. The thickest layer in a typical artery is the
   a. tunica interna. b. tunica media. c. tunica externa. d. endothelium. e. elastic layer.

2. Blood flow through the capillaries is determined by the
   a. vennule. b. elastic arteries. c. muscular arteries. d. precapillary sphincters. e. endothelium.

3. At any given moment, the bulk of the blood volume is located in the
   a. arteries. b. capillaries. c. veins. d. sinuses. e. fenestrated capillary beds.

4. The aortic arch gives rise to the
   a. brachiocephalic artery. b. left subclavian. c. left common carotid. d. all of the above e. none of the above

5. The small intestine is supplied with blood by the
   a. inferior mesenteric. b. celiac. c. superior mesenteric. d. circumflex. e. popiteal.

6. The anastomosis of the superior mesenteric vein and the splenic vein form the
   a. hepatic portal vein. b. hepatic vein. c. inferior mesenteric vein.
   d. brachiocephalic vein. e. inferior vena cava.

7. The major vein draining the brain is the
   a. superior mesenteric vein. b. external jugular. c. internal jugular. d. internal carotid. e. basilar.

8. The correct sequence for a drop of blood passing to the foot would be
   a. femoral - common iliac - external iliac - popiteal - tibial. b. external iliac - common iliac - femoral - popiteal - tibial. c. common iliac - external iliac - popiteal - femoral - tibial. d. common iliac - external iliac - femoral - tibial - popiteal.
   e. common iliac - external iliac - femoral - popiteal - tibial.

9. The difference between the systolic and diastolic pressures is the
   a. mean arterial pressure. b. average arterial pressure. c. venous pressure.
   d. pulse pressure. e. none of the above

10. Dividing the mean arterial pressure by the total peripheral resistance will yield the
    a. pulse pressure. b. systolic pressure. c. diastolic pressure. d. cardiac output. e. amount of vasodilation.

11. Blood pressure may be increased by
    a. increasing cardiac output. b. increasing peripheral resistance. c. release of norepinephrine. d. release of ADH. e. more than one of the above is correct

12. Increasing baroreceptor activity will
    a. increase vasomotor activity. b. decrease cardioinhibitory activity. c. increase cardioacceleratory activity. d. increase peripheral resistance. e. decrease blood pressure.

13. Which of the following hormones will decrease blood pressure?

a. ADH   b. norepinephrine   c. epineprhine   d. aldosterone   e. ANF

14. Fluid moves from the tissues into the capillaries when

a. hydrostatic pressure exceeds osmotic pressure. b. osmotic pressure exceeds hydrostatic pressure. c. hydrostatic pressure equals osmotic pressure. d. blood pressure is high. e. osmotic pressure is low.

15. A regular program of physical exercise causes the heart to

a. become larger. b. become more efficient. c. increase the resting stroke volume. d. decrease the resting heart rate. e. more than one of the above is correct

CHAPTER 22
The Lymphatic System and Immunity

OVERVIEW

The lymphatic system is very closely associated with the cardiovascular system. In certain respects it is anatomically similar, consisting of a large number of vessels that conduct fluid. These vessels pick up tissue fluid and eventually return it to the general circulation, and in this respect, the lymphatic system serves as an adjunct to the cardiovascular system. The lymphatic system is more than a simple fluid return system. There are associated lymphoid organs that contain large numbers of macrophages, large, phagocytic white cells. These macrophage systems survey and cleanse the tissue fluid of foreign material and debri before it again enters the circulation. In addition, many of these lymphoid organs generate lymphocytes. It is the lymphocytes working in conjunction with the macrophages which are responsible for the body's immune response. The lymphatic system is the center of the immune response and this is a second major function of this system. Finally the lymphatic system maintains a dynamic osmotic balance with the interstitial fluid. In this way it prevents local variation in tissue fluid composition.

This chapter begins with a consideration of the anatomy of the lymphatic system. This includes a survey of the lymph vessels, the lymphocytes, the lymphatic tissues, and the lymphoid organs. Once the anatomy has been established you will proceed to a consideration of the role of the lymphatic system in body defenses. First to be considered will be the non-specific defenses and then the world of specific immunity will be examined.

When reading this chapter, especially the sections dealing with the immune response, you should be aware of the fact that this area is one of the most exciting frontiers in biological research today. New and significant discoveries are being made almost daily in this area and these discoveries promise to revolutionize a number of areas in health and medicine in the future.

CHAPTER OUTLINE

A. Introduction
B. Anatomy of the lymphatic system
   1. Lymphatic vessels
   2. The lymphocytes
      a. The lifespan and circulation of lymphocytes
      b. Lymphopoiesis
   3. Lymphatic tissues
   4. Lymphatic organs
      a. The thymus
      b. Lymph nodes
      c. The spleen
   5. Integration with the immune system
C. Nonspecific defenses
   1. Physical barriers
   2. Phagocytic cells
      a. Orientation and phagocytosis
   3. Immunological surveillance
   4. Complement
   5. Inflammation
D. Specific immunity
   1. Humoral immunity
      a. B cells and antibody production
      b. Antibody structure and function
      c. Antibody diversity
   2. Cellular immunity
   3. The regulation of the immune response
      a. Chemical communication and coordination
E. Patterns of immune response
   1. Immunological competence
      a. Tolerance
   2. Stress and the immune response
F. Clinical patterns
   1. Immune deficiency diseases
   2. Disorders of excessive antibody production
      a. Immune complex diseases
      b. Autoimmune disease
   3. Immunosuppression

LEARNING ACTIVITIES

Complete the following items by supplying the appropriate word or phrase.

1. The fluid found inside of the lymphatic vessels is known as _____ and is similar in composition to interstitial fluid.

2. List the the three major functions of the lymphatic system.

   _____

   _____

   _____

3. Lymphatic capillaries have one end _____.

4. The two lymphatic ducts that eventually join the circulatory system are the _____ duct on the left side and the _____ duct on the right side.

5. Blockage of the lymphatic drainage from a limb produces the condition _____, in which lymph accumulation greatly distends and distorts the limb.

6. The lymphocytes of the body form three distinct populations, the _____, the _____, and the _____.

7. Approximately 80% of the lymphocytes are _____.

8. The T cell population that attack foreign or infected cells are the _____ T cells.

9. _____ T cells regulate and coordinate the immune response.

10. _____ produce antibodies.

11. _____ cells attack foreign cells, normal cells infected with viruses, and cancer cells.

12. Eighty percent of the lymphocytes will live for at least _____ years, and some for twenty years.

13. The process of lymphocyte generation is known as _____.

14. Lymphocyte production begins with the hemocytoblast of the bone marrow which produces _____ different populations of cells.

15. The group of prolymphocytes that remains in the bone marrow become _____ and _____

16. The second group of cells migrates from the bone marrow to the thymus gland where the hormone _____ converts them into T cells.

17. A _____ is a dense aggregation of lymphocytes in an area of loose connective tissue.

18. Lymphatic nodules are usually found beneath an _____.

19. The large nodules found in the wall of the pharynx are the _____.

20. _____ patches are clusters of nodules beneath the epithelium of the small intestine.

21. The _____ contains an extensive mass of fused lymphatic nodules.

22. The three major lymph organs are the _____, the _____, and the _____.

23. The _____ begins to involute after puberty but seems to remain functional throughout life.

24. The lymph node is surrounded by a _____ capsule.

25. _____, or inward extensions of the lymph node capsule form a series of incomplete partitions.

26. The cortex of the lymph node is dominated by _____ cells and the medulla region by _____.

27. Label figure 22.1, the structure of a lymph node, using the terms listed below. Color the various regions and structures. Refer to figure 22.8 in your textbook for assistance.

efferent lymphatic vessel	afferent lymphatic vessel	cortex
trabecula	capsule	medullary cord
medulla	hilus	

Label: efferent lymphatic vessel   afferent lymphatic vessel   cortex   trabecula   capsule
        medullary cord   medulla   hilus

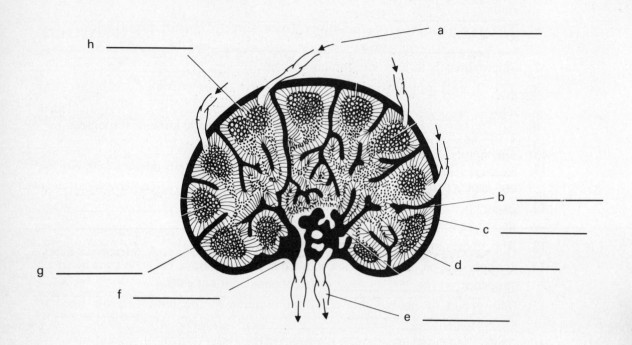

Figure 22.1

28. The lymphocytes of the lymph nodes execute immune reactions and the
    _____ engulf foreign materials.

29. _____ is excessive enlargement of lymph nodes.

30. In the adult, the largest collection of lymphoid tissue in the body is found in the
    _____.

31. Spleen functions include the _____ of abnormal blood components
    and the initiation of _____ responses.

32. The _____ pulp of the spleen contains the bulk of the lymphocytes.

33. The lymphatic system is one part of a more diffuse _____ system.

34. The defense systems of the body are divided into the _____ defenses
    and _____.

35. The body is protected from entry by foreign agents by the intact skin and the
    _____ that lines all of the organs that open to the outside.

36. The cellular first line of defense against pathogens are the _____
    cells.

37. The two general groups of phagocytic cells are the _____ and the
    _____.

38. The microphages include the _____ and _____.

39. The fixed _____ are permanent residents of organs and tissues and
    have no ability to move.

40. The fixed macrophages are usually found scattered throughout
    _____ tissues.

41. The fixed macrophages of the CNS are the _____, those found in the
    liver are the _____, and those of the skin are the _____.

42. The _____ can move freely around the tissues.

43. Both free and fixed macrophages are derived from the _____.

44. Both microphages and macrophages can leave the circulation by a process called
    _____.

45. The monitoring of normal tissues by natural killer cells is the non-specific defense
    function known as _____ surveillance.

46. NK cells can recognize cancerous cells and virus infected cells by the presence of
    new _____ which appear on the cell membrane.

47. The complement system consists of _____ plasma proteins.

48. Once activated the complement system will execute a number of defense measures.
    These include

    a. _____

    b. _____

    c. _____

    d. _____

    e. _____

49. Complement is most effective when the complement proteins bind to
    _____ molecules.

50. The complement factor _____ can initiate and sustain complement
    reactions in the absence of antibody.

51. The most comprehensive and complex nonspecific response to connective tissue in-
    jury is _____.

52. _____ are molecules capable of stimulating an immune response.

53. Immunity that is mediated by B lymphocytes and antibody is known as
    _____ immunity.

54. Antibodies do not target whole antigens, but certain parts which are the
    _____ sites.

55. To function as an antigen, a molecule must have at least _____ an-
    tigenic determinant sites.

56. _____ can become antigenic by combining with a body protein
    which serves as a carrier molecule.

57. B cells that have been exposed to antigen differentiate into _____ cells
    and _____.

58. The primary response to antigen in humoral immunity is the production of
    _____ by plasma cells.

59. It takes about 1 to 2 weeks following exposure to antigen for the antibody
    _____ to reach peak levels.

60. The secondary or anamnestic response occurs when _____ are exposed to antigen a second time.

61. A vaccine functions by promoting a _____ response in an individual to a given antigen and thereby creating memory cells which will function in any subsequent exposure.

62. An antibody molecule consists of a total of _____ polypeptide chains.

Amino Acid

63. Each polypeptide chain of an antibody contains a constant and a _____ region of amino acids.

64. It is the _____ region that is antigen specific.

65. Activation of antibody can result in about five different processes. These processes are as follows.

    a. _____
    b. _____
    c. _____
    d. _____
    e. _____

66. For each statement below, write the appropriate immunoglobulin class in the space provided. Abbreviate the class as IgA, IgG, IgD, IgE, and IgM.

    _____ This immunoglobulin is bound the the B cell membrane and serves as the antigen receptor.

    _____ It is the most abundant of all of the immunoglobulins.

    _____ It is the immunoglobulin that is responsible for the agglutination of cross matched blood.

    _____ This immunoglobulin attaches to the membrane of Mast cells and basophils.

    _____ Members of this class are found in glandular secretions.

67. Cell mediated immunity is the function of the _____ lymphocytes.

68. When T cells are exposed to antigen they will divided to form memory cells and _____ cells.

69. Cytotoxic cells will destroy any cell that contains the _____ that sensitized them.

70. The special series of glycoproteins found in the membrane of all nucleated cells, and which function is antigen recognition, are the _____.

71. There are two classes of HLA molecules, _____ are found on all nucleated cells while _____ are found only on lymphocytes and macrophages.

72. Lymphocytes can recognize antigens which are attached to the body's own genetically determined _____ or they can recognize foreign (non-self) _____ and initiate an immune response against them.

73. Below is the sequence of events involved in the activation and regulation of an immune response. Complete each statement by writing in the appropriate word.

    Before an antigen can stimulate any immune response it must be first processed by an _____a_____ cell, usually a macrophage.

    The processed antigen is then bound to the macrophage's _____b_____.

    The bound and processed antigen is now presented to a _____c_____ cell.

    If the antigens are bound only to the macrophage's class I HLA then _____d_____ T cells will be activated, and a cell mediated response will be initiated.

If the antigen is bound to both class I and II HLA then T _____e_____ cells will be activated, and a humoral immune response will be initiated.

_____f_____ T cells are activated by contact with antigen-class I HLA complex while _____g_____ T cells are activated by contact with antigen-class II HLA complex.

In order for B cells to become active plasma cells, they must be presented antigen by the _____h_____ T cells.

_____i_____ T cells regulate the magnitude of the response by inhibiting B cell activation.

a. _____

b. _____

c. _____

d. _____

e. _____

f. _____

g. _____

h. _____

i. _____

74. The specific and nonspecific defenses of the body are coordinated by a series of chemical signals released by the defense cells. _____ are released by the monocyte-macrophage system while _____ are released by the lymphocytes.

75. _____ make T cells more sensitive to antigens and accelerate the rate of production of both cytotoxic and regulatory T cells.

76. In addition to the effects which the interleukins have on the immune cells they also have the following effects.

a. _____

b. _____

c. _____

d. _____

e. _____

77. Gamma _____ which is released by T cells stimulates macrophage and NK activation.

78. _____ and _____ are released by helper T cells and activate B cells.

79. _____ which is secreted by activated macrophages will slow tumor growth.

80. Below is a list of events that occur in a defensive response of the body. List them in their proper sequence of occurrence by placing a number in front of each statement. For example, the first event to occur in the general defense response will have the number 1 placed in front of it.

_____ B cells are activated to form plasma cells.

_____ Inflammation, monokines, and lymphokines draw in large numbers of phagocytes.

_____ NK cells invade the area and destroy foreign cells.

_____ Cytotoxic T cells invade the area.

81. The fetus has passive immunity based upon IgG derived from the _____.

82. The ignoring of our own antigens by our immune systems is known as _____.

83. Chronic stress results in _____ of the immune system.

84. Match the clinical term listed below with the appropriate descriptive statement which follows.

a. SCID   b. AIDS   c. ARC   d. Kaposi's sarcoma   e. ELISA test   f. anaphylaxis
g. hives   h. immune complex disease   i. autoimmune disease   j. cyclosporin A

_____ results when the immune system fails to tolerate self antigens

_____ result from the deposition of antigen-antibody complexes in the blood vessels

_____ blood test used to screen blood donors for AIDS

_____ causes destruction of the T cells and collapse of the immune system

_____ suppresses the immune response by stimulating suppressor T cells

_____ rare form of skin cancer found mostly in AIDS victims

_____ a chronic but mild condition characterized by lymphadenopathy

_____ condition where an individual fails to develop either cellular or humoral immunity

_____ allergic response in which raised welts appear on the surface of the skin

SELF TEST

Circle the correct answer to each of the following questions.

1. The capillaries of the lymphatic system are
   a. open at both ends.   b. closed at one end.   c. thicker than the capillaries of the circulation.   d. more extensive than the circulatory capillaries.   e. none of the above.

2. The thoracic duct enters the
   a. right subclavian vein.   b. right internal jugular vein.   c. left subclavian vein.
   d. left internal jugular vein.   e. left external jugular vein.

3. Which of the following are lymphocytes?
   a. T cells   b. B cells   c. NK   d. plasma cells   e. more than one of the above is correct.

4. The cells which are responsible for immunological surveillance are the
   a. B cells.   b. macrophages.   c. T cells.   d. NK cells.   e. neutrophils.

5. Peyer's patches are aggregations of lymphoid tissue found in the
   a. wall of the small intestine.   b. appendix.   c. pharynx.   d. axillary region.
   e. groin region.

6. The cortical zone of the lymph node is dominated by
   a. NK cells.   b. connective tissue.   c. B cells.   d. T cells.   e. neutrophils.

7. A major function of the spleen is
   a. generation of T cells.   b. generation of NK cells.   c. cleansing of the blood.
   d. formation of B cells.   e. cleansing of tissue fluid.

8. Which of the following is not a nonspecific defense mechanism?

   a. inflammation   b. physical barriers   c. immunity   d. phagocytosis   e. immunological surveillance

9. The _____ defense system consists of eleven plasma proteins.

   a. inflammatory   b. humoral immunity   c. complement   d. lymphoid   e. immunological surveillance

10. Haptens or partial antigens may behave as a full antigen after they bind with

    a. a lymphocyte.   b. a macrophage.   c. a body protein.   d. another antigen.
    e. none of the above.

11. The antibody which functions as the antigen receptor for B cells is

    a. IgA.   b. IgD.   c. IgG.   d. IgM   e. IgE

12. Before an antigen can stimulate a lymphocyte, it must first be processed by a

    a. macrophage.   b. NK.   c. cytotoxic T cell.   d. neutrophil.   e. microphage

13. The antigen on macrophages is bound to the

    a. nucleus.   b. lysosomes.   c. ribosomes.   d. HLA.   e. centriole.

14. Before B cells can respond to antigen, they must receive a signal from

    a. cytotoxic T cells.   b. suppressor T cells.   c. helper T cells.   d. plasma cells.
    e. macrophages.

15. The lymphokine that stimulates fever and activates T cells is

    a. interferon.   b. BCDF.   c. BCGF.   d. interleukin.   e. TNF.

CHAPTER 23
The Respiratory System

OVERVIEW

The respiratory system is one of those systems that are closely associated with the circulatory system. The circulatory, respiratory, excretory-urinary, and digestive systems are sometimes referred to collectively as the homeostatic systems because they all play a direct role in maintaining the constant fluid environment required by cells. The respiratory system's contribution to this process is the placement of oxygen into the blood and the removal of carbon dioxide from the blood. The respiratory system therefore controls respiratory gas homeostasis.

This chapter follows the same organizational pattern as the previous ones. The functional anatomy of the respiratory system is the first consideration. Beginning with the nose, the anatomy of the system is traced through the pharynx, larynx, trachea, and into the lungs where gas exchange with the blood occurs. Once the major anatomical features have been established, an examination of the system's physiology is undertaken. This begins with the process of respiration, or the forces and processes involved with the movement of gases into and out of the lungs. You will see that some of the same principles that applied to blood flow, pressure, resistance, and flow, also apply to the movement of gases in the respiratory system. Once the mechanisms of gas movement have been discussed the topic of external respiration, the actual exchange of gases between the lung and blood is considered. This is followed by the mechanisms by which the blood actually transports gases to the tissues. Finally, the topic of respiratory control is discussed. Here you will examine the mechanisms by which the respiratory rate is regulated in order to meet the various demands of the body.

CHAPTER OUTLINE

LEARNING ACTIVITIES

Complete each of the following items by supplying the appropriate word or phrase.

1. List the five primary functions of the respiratory system.

   a. _____

   b. _____

   c. _____

   d. _____

   e. _____

2. The conducting passageways that deliver air to the lungs comprises the _____ respiratory tract while the exchange surfaces and delicate passageways are contained in the _____ respiratory tract.

3. Air enters the respiratory system via the external _____.

4. The _____ is that part of the nasal cavity contained within the flexible tissues of the external nose.

5. The nasal _____ divides the nasal cavity into left and right halves.

6. The floor of the nasal cavity is formed by the _____ palate.

7. The _____ is an extension of the hard palate that separates the superior nasopharynx from the rest of the pharynx.

8. The internal _____ are the connections between the nasal cavity and the nasopharynx.

9. The nasal _____ are inward projections of the lateral walls of the nasal cavity.

10. The conchae provide a passageway for air that results in a _____ of dust particles, pollen, microorganisms, and other debri from the air.

11. The respiratory epithelium consists of a _____, ciliated, columnar epithelium with numerous goblet cells.

12. The cilia and mucous of the respiratory membrane sweep trapped particles towards the pharynx where they will be _____.

13. The respiratory epithelium plus the lamina propria it rests upon are termed the _____.

14. In addition to cleansing, the nasal cavity warms and _____ air before it enters the lower tract.

15. The _____ is a chamber shared by both the respiratory and digestive tracts.

16. The three regions of the pharynx are the _____, _____, and _____.

17. The oropharynx and laryngopharynx are lined by a _____ epithelium that can resist abrasion.

18. Match the terms concerning the larynx below with the appropriate descriptive statements which follow.

    a. glottis b. thyroid cartilage c. cricoid cartilage d. arytenoid cartilages e. corniculate cartilages f. cuneiform cartilage g. epiglottis h. intrinsic ligaments
    i. extrinsic ligaments j. ventricular folds k. true vocal cords

    _____ cartilage that the thyroid cartilage sits on top of

    _____ blade shaped cartilage that projects into the pharynx above the glottis

    _____ the entrance into the larynx

    _____ cartilage that forms most of the anterior surface of the larynx

_____ the false vocal cords

~~Omit~~ _____ attach the thyroid cartilage to the hyoid bone

~~Omit~~ _____ cartilages that articulate with the superior border of the cricoid cartilage

_____ highly elastic folds that produce sound

~~Omit~~ _____ along with the arytenoid cartilages these are involved with the production of sound

~~Omit~~ _____ bind the articulate surfaces of the laryngeal cartilages together

~~Omit~~ _____ elongate, curving, wedge shaped cartilages

19. Sound is produced when _____ passing over the vocal chords causes them to vibrate.

20. The pitch of the sound produced by the vocal chords is dependent upon the length of the chord and the amount of _____ placed upon the chord.

21. The _____ laryngeal muscles regulate the tension on the vocal chords and open and close the glottis.

22. The _____ laryngeal muscles position and stabilize the larynx.

23. Laryngeal _____ are reflexes that restrict the entrance of potentially dangerous materials into the lower tract.

24. Inflammation of the larynx leads to _____ in which the voice may be hoarse or disappear completely.

25. The _____ is the tube that descends from the larynx.

26. The _____ of the trachea is similar to that lining the nasal cavity.

27. The wall of the trachea is reinforced by c-shaped rings of _____ which prevent collapse of the trachea. In this regard, the trachea resembles the hose of a vacuum cleaner, flexible, but not subject to collapsing.

28. The trachea branches to give rise to the _____ bronchi.

29. The right lung has _____ lobes and the left lung has _____ lobes.

30. Label figure 23.1, the respiratory tract, using the terms listed below. Color each part. Refer to figure 23.1 in your textbook for assistance.

frontal air sinus	sphenoid air sinus	middle concha	right lung
inferior concha	superior concha	hard palate	soft palate
pharynx	left primary bronchus	larynx	esophagus
epiglottis	right primary bronchus	left lung	trachea

31. The parenchyma of the lungs is rich in _____ fibers which give the lungs their great distensibility.

32. The primary bronchi give rise to the __a__ bronchi, one for each lobe of the lungs. Once in the lobes, the bronchial tubes keep dividing until finally the __b_ are reached. Each of these supplies one lobule of the lung. Within the lobule these tubules continue to divide, forming the __c_ bronchioles which have a diameter of less than 0.5 mm. The epithelium of these small bronchioles is now __d_, and cilia as well as goblet cells are rare. Each of these bronchioles gives rise to several __e_ bronchioles which supply the respiratory surface of the lungs.

a. _____

b. _____

c. _____

d. _____

e. _____

Label: frontal air sinus   sphenoid air sinus   middle concha   right lung   inferior concha
        superior concha   hard palate   soft palate   pharynx   epiglottis   larynx
        esophagus   left primary bronchus   larynx   esophagus   epiglottis
        right primary bronchus   left lung   trachea

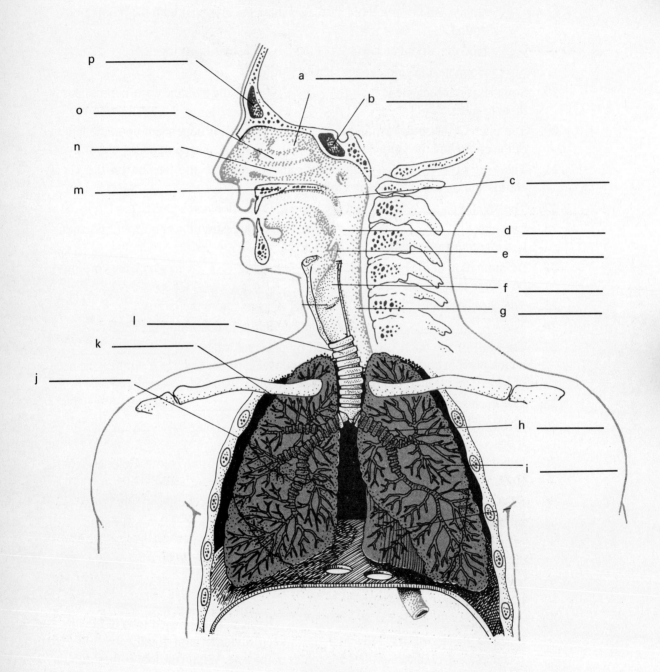

Figure 23.1

33. The respiratory bronchioles open into chambers called the _____ ducts,
     and these have blind pockets termed _____ where gas exchange occurs.

34. Label figure 23.2, the alveoli and bronchioles, using the terms listed below. Color
     each part. Refer to figure 23.9 in your textbook for assistance.

     alveolus   alveolar duct   terminal bronchiole
     respiratory bronchiole   alveolar capillary network   pulmonary venule
     pulmonary arteriole   alveolar sacs

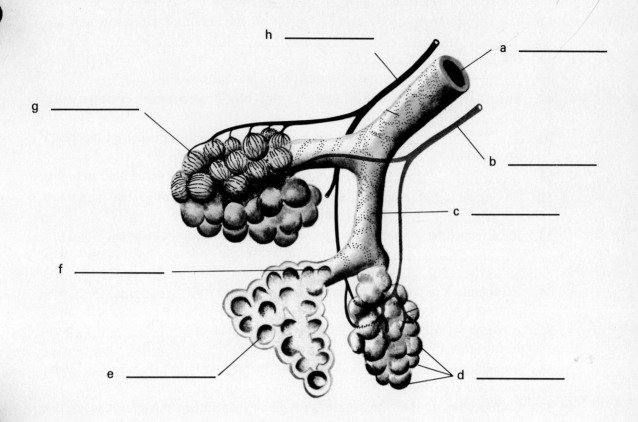

h _____

a _____

g _____

b _____

c _____

f _____

e _____

d _____

Figure 23.2

35. The alveolar epithelium is a simple _____ epithelium.

36. Septal cells produce an oily substance, _____, which coats the epithelium and prevents the aveoli from collapsing.

37. Alveolar _____ patrol the aveolar epithelium and phagocytize any particles small enough to get past the defenses of the upper tract.

38. Gas exchange occurs in areas where the _____ which underlies the aveolar epithelium has fused with that of the capillary endothelium.

39. All of the cilia that line the lower respiratory tract beat in a direction which moves mucous upward toward the pharynx and form the _____ escalator which removes trapped foreign particles.

40. Each lung lies in a _____ cavity.

41. It is the _____ that divides the pleural cavity in half.

42. The _____ pleural membrane lines the cavity and the _____ covers the lung and mediastinum.

43. The pleural cavity actually represents a _____ space as the opposing parietal and visceral pleura layers are usually in close contact with only a thin layer of lubricating fluid separating them.

44. Inflammation of the pleura leads to the condition of _____.

45. Prior to birth, the lungs of the fetus are _____.

46. The first breath following birth not only pulls airs into the lungs, but the drop in pressure pulls blood into the _____ vessels.

47. The respiratory process can be broken down into four phases which are _____, _____, _____, and _____.

48. _____ is the movement of air into and out of the lungs.

49. The diffusion of gases between the blood and alveoli constitute _____ respiration.

50. _____ respiration is the exchange of gases between the blood and tissue fluid.

51. _____ is the exchange of gases between the tissue fluid and cells.

52. The major difference between gases and liquids as fluids is that gases are _____ while liquids are not.

53. If you reduce the volume which a gas occupies the pressure within the enclosure will _____, and if you increase the volume, the pressure will _____.

54. If a volume of gas is enclosed, raising its temperature will _____ the pressure which is exerts.

55. Air (or any gas) under pressure, tends to flow from areas of _____ pressure to areas of _____ pressure.

56. Inspiration will not occur unless intrapulmonary pressure is _____ than atmospheric pressure.

57. Contraction of the diaphragm and other inspiratory muscles causes the _____a_____ cavity to increase in volume. This results in a _____b_____ in intraplural pressure. Decreasing intrapleural pressure results in a drop in _____c_____ pressure. When the intrapulmonary pressure is less than that of the _____d_____ air will fill the lungs. Relaxation of the inspiratory muscles results in the pleural cavity _____e_____ in volume. This causes an increase in intrapleural pressure. The increase in intrapleural pressure along with the _____f_____ recoil of the lungs causes _____g_____ pressure to exceed that of the atmosphere and air moves out.

    a. _____
    b. _____
    c. _____
    d. _____
    e. _____
    f. _____
    g. _____

58. During maximum ventilation, intrapulmonary pressure may be _____ less than atmospheric pressure during inspiration and _____ greater during expiration.

59. An injury that allows air into the pleural cavity is known as a _____ and can result in a collapsed lung.

60. The diaphragm and _____ intercostals are the major respiratory muscles for inspiration.

61. Forced expiration is brought about by the _____ intercostals.

62. The lungs and chest wall return to resting volume following relaxation of the inspiratory muscles because of _____.

63. Normal quiet breathing is termed _____ while forced breathing is known as _____.

64. Mouth-to-mouth resuscitation provides ventilation to a victim who has stopped breathing but CPR restores _____ and ventilation.

65. Match the term listed below with the appropriate descriptive phrase that follows.

a. tidal volume   b. expiratory reserve   c. inspiratory reserve   d. vital capacity   e. residual volume   f. minimum volume   g. respiratory rate   h. respiratory minute volume

_____ number of breaths per minute

_____ volume of air that moves during a normal quiet breathing cycle

_____ total volume of air that can be moved and is equal to the TV + IR + ER

_____ volume of air that always remains in the lungs

_____ can be thought of as being the respiratory equivalent of cardiac output because it is equal to the tidal volume times the respiratory rate

_____ respiratory volume following collapse of the lungs

_____ total volume of air that can be forcefully expired following a normal expiration

_____ total volume of air that can be forcefully inspired following a normal inspiration

66. Increasing pressure will increase the number of molecules of a gas that will dissolve into _____.

67. The ease with which gas molecules go into solution are a measure of the gases _____.

68. _____ law relates the number of gas molecules in solution to pressure and solubility.

69. Gas X is highly soluble in water but gas Z is not. It might be possible to get equal concentrations of X and Z in solution by increasing the _____ of Z.

70. The principle that in a mixture of gases the pressure exerted by an individual gas is equivalent to its percentage of the total is known as _____.

71. The pressure exerted by a single gas in a mixture is known as its _____ pressure.

72. Five different gases are exerting a pressure of 1000 mm Hg. Gas X makes up 10% of the total. Its partial pressure would be _____ mm Hg.

73. The volume of the respiratory tract that does not consist of the aveoli is known as the anatomic _____ space.

74. The diffusion of a particular gas from air into solution depends soley upon the _____ of that gas.

75. Carbon dioxide diffuses into the alveoli from the pulmonary capillaries because the partial pressure of carbon dioxide is _____ in the capillaries than in the alveoli.

76. Oxygen diffuses from the alveoli into the blood because the _____ pressure of oxygen is greater in the alveoli than in the blood.

77. At the level of the tissues, oxygen diffuses from the blood because the partial pressure of oxygen in the blood is _____ than it is in the tissues.

78. About _____ of oxygen is transported in solution in the plasma.

79. _____ has a high affinity for oxygen and at the oxygen partial pressure found in alveolar air it is about 97.5% saturated.

must
know

80. The steep initial slope of the __Oxy hemoglobin__ saturation curve shows that a small drop in oxygen levels in the tissue sill result in a large release of oxygen by hemoglobin. *will*

81. The flattened portion at the top of the oxygen-hemoglobin saturation curve is significant because it means that hemoglobin can load up with oxygen under a wide range of atmospheric __Oxygen__ levels.

82. If the pH of the blood begins to decline, hemoglobin will release its oxygen at a __higher__ oxygen partial pressure in the tissues.

83. The reason for this greater release of oxygen under acid conditions is that the __hydrogen ion__ tends to push oxygen off of the hemoglobin molecule.

84. Increasing temperature causes hemoglobin to __Release__ oxygen.

85. Elevating __DPG__ in red blood cells will cause hemoglobin to release about 10 percent more oxygen at a given partial pressure of oxygen than it would do otherwise.

86. One reason that maternal hemoglobin can pass oxygen to fetal hemoglobin is that fetal hemoglobin has a much __higher__ affinity for oxygen than does adult hemoglobin.

87. In addition to the hemoglobin differences, the fetus has a __higher__ hematocrit, and about __50__ percent more hemoglobin per unit of blood.

88. Carbon dioxide is transported in __Plasma__, bound to __hemoglobin__, and in the form of __Bicarbonate__ *that is*

89. About __1__ percent of total carbon dioxide is transported dissolved in the plasma.

90. About 23 percent of carbon dioxide combines with hemoglobin to form __Carbinohemoglobin__

91. __70__ percent of total carbon dioxide is transported in the form of bicarbonate.

92. The reaction of water with carbon dioxide to form carbonic acid occurs inside of the red blood and is catalyzed by the enzyme __Carbonic Anhydrase__

93. Carbonic acid dissociates to form hydrogen ion and __Bicarbonate__

94. The hydrogen ion formed by dissociation of carbonic acid combines with hemoglobin and the bicarbonate diffuses into the plasma where it associates with __Sodium__.

95. In order to maintain electrical neutrality, every time a negative bicarbonate diffuses out of a red cell, a negative __Chloride__ moves into the cell.

96. Respiration is under the control of three pairs of nuclei located in the __a__ formation of the pons and medulla. *Reticular* The site of initiation is the __b__ which contains both *Respiratory* an inspiratory and expiratory center. During quiet breathing respiration is initiated by impulses from the inspiratory center to the __c__ muscles. *Rhythmic Center* When the inspiratory *Inspiration* center shuts down the muscles relax and __d__ occurs passively. The expiratory center is active only during forced breathing. The expiratory center is activated by the *Expiration* __e__ and then stimulates the expiratory muscles. Respiratory rate and depth can be modified by the pons centers. The __f__ center inhibits both the inspiratory center *Pneumotaxic* and apneustic center thereby preventing extended inspiration. A decrease in pneumotaxic activity permits the apneustic center to become more active and breathing rate drops but __g__ of inspiration increases. *depth*

a. _____

b. _____

c. _____

d. _____

e. _____

f. _____

g. _____

97. When the lungs reach a critical amount of stretch during inflation, receptors become active and send inhibitory signals to the _____ center.

98. The _____ reflex has receptors that activate the inspiratory center when the lungs are collapsing near the end of a forced expiration.

99. The inflation and deflation reflexes constitute the _____ reflex.

100. The Hering-Breuer reflex is significant only during _____ breathing.

101. The respiratory centers are sensitive to the chemoreceptors known as the _____ and _____ bodies which monitor arterial oxygen and carbon dioxide levels.

102. The inspiratory center is especially sensitive to _____ ions in the CSF which are formed by the reaction of carbon dioxide with water.

103. It is _____, not oxygen, that determines respiratory rate.

104. Hypercapnia will result in immediate _____.

105. Carbon dioxide has the greatest effect on the respiratory centers because ordinarily, _____ in arterial blood is never low enough to activate receptors.

106. A condition in which infants 2 - 4 months old suddenly stop breathing and die is known as ___SIDS___ or "crib death."

107. Match the clinical condition listed below with the appropriate descriptive statement which follows.

a. cystic fibrosis   b. emphysema   c. COPD   d. pneumonia   e. pulmonary thromboembolism   f. respiratory acidosis   g. carbon monoxide   h. bronchitis   i. tuberculosis   j. lung cancer

_____ caused by accumulating carbon dioxide, it alters protein structure and function throughout the body

_____ bacterial infection of the lungs which results in inflammatory reactions that alter and distort the respiratory passageways

_____ disease caused by chronic irritation in which the alveoli expand and their wall become infiltrated with fibrous connective tissue

_____ often results in an expansion of the chest in the latter stages of emphysema.

_____ blocking of a pulmonary blood vessel by a thrombus

_____ causes a severe poisoning by competing with oxygen for hemoglobin

_____ inflammatory disease of the lobules usually caused by bacteria, especially after body defenses have been depressed

_____ 85 to 90 percent of all cases of this disease are the direct result of smoking cigarettes

_____ inherited disease characterized by excessive production of viscous mucus

_____ an inflammation of the bronchial lining usually caused by irritation

SELF TEST

Circle the correct answer to the following questions.

1. Air moves from the nasal cavity to the nasopharynx via the

    a. oropharynx.   b. external nares.   c. uvula.   d. internal nares.   e. eustachean tubes.

2. The nasal cavity is lined by

    a. stratified squamous epithelium.   b. pseudostratified, ciliated, columnar epithelium.   c. cuboidal epithelium.   d. simple squamous epithelium.   e. none of the above.

3. Which of the following is not a laryngeal cartilage?

    a. cricoid   b. thyroid   c. arytenoid   d. corniculate   e. costal

4. The right lung has _____ lobes.

    a. 1   b. 2   c. 3   d. 4   e. 5

5. The alveolar ducts receive air directly from the

    a. alveoli.   b. terminal bronchioles.   c. respiratory bronchioles.   d. primary bronchi.   e. tertiary bronchi.

6. Surfactant is produced by the

    a. aveolar ducts.   b. respiratory bronchioles.   c. septal cells.   d. aveolar macrophages.   e. more than one of the above is correct.

7. The exchange of gases between the pulmonary capillaries and the alveoli is known as

    a. external respiration.   b. internal respiration.   c. cellular respiration.   d. pulmonary respiration.   e. ventilation.

8. Air will move into the lungs when the

    a. intrapleural pressure is greater than the intrapulmonary pressure.   b. intrapulmonary pressure is greater than the intrapleural pressure.   c. atmospheric pressure is less than intrapulmonary pressure.   d. atmospheric pressure is less than intrapleural pressure.   e. atmospheric pressure is greater than intrapulmonary pressure.

9. When a gas is enclosed in a smaller volume, the pressure it exerts

    a. goes down.   b. goes up.   c. does not change.

10. The total volume of air that a person can move is their

    a. inspiratory reserve.   b. tidal volume.   c. expiratory reserve.   d. vital capacity.   e. dead air space.

11. If the partial pressure of a gas in the pulmonary circulation is 200 mm Hg and its partial pressure in the alveolus is 100 mm Hg, the gas will move

    a. from the pulmonary circulation to the alveolus.   b. from the alveolus to the pulmonary circulation.   c. either way depending upon what the partial pressures of the other gases are.   d. it will not move at all because gases cannot diffuse.   e. none of the above

12. The conversion of carbon dioxide into bicarbonate occurs predominantly in the red blood cell because

    a. carbon dioxide is generated by the RBC.   b. the enzyme carbonic anhydrase is inside of the RBC.   c. free water is inside of the RBC.   d. potassium in inside of the RBC.   e. chloride is in the plasma.

13. Which of the following will cause hemoglobin to give up more oxygen?
   a. acid pH   b. increased temperature   c. DPG   d. reduced tissue fluid oxygen partial pressure.   e. more than one of the above is correct

14. The _____C_____ center causes sustained inspiratory movements.
   a. inspiratory   b. expiratory   c. apneustic   d. pneumotaxic   e. medullary

15. The major factor regulating respiratory rate is
   a. tissue oxygen levels.   b. arterial oxygen levels.   c. carbon dioxide levels.   d. the pneumotaxic center.   e. the apneustic center.

CHAPTER 24
The Digestive System

OVERVIEW

This is the second system that works in close conjunction with the circulation to maintain homeostasis for the cells. The respiratory system contributed by maintaining proper respiratory gas levels. The digestive system contributes by maintaining proper levels of nutrients, both organic and inorganic. The digestive system is essentially a long tube that is open at both ends. Food enters one end of the system and moves slowly through the tube being both mechanically and chemically processed as it moves. Along the way several accessory glands outside of the tube contribute to the chemical processing. The processed nutrients are absorbed across the mucosal lining of the tract into the circulation and lymphatics. Eventually the undigested food materials are eliminated from the far end of the tube.

Besides the breakdown and absorption of nutrients, the digestive tract also contributes to the excretory function. Metabolic waste products, including the breakdown products of hemoglobin, are eliminated from the body via the liver bile. It should be pointed out that the elimination of undigested food materials is not considered excretion because these materials have never actually been in the body proper. We only excrete materials which have been in the blood at one time and therefore the materials contained in liver bile qualify as excreted products but those that remain in the digestive tract do not.

This chapter approaches the digestive tract by first examining its functional anatomy. As the wall of the digestive tube is the functional part of the system, a discussion of the common wall structure of the tract is the first topic covered. Then the functional anatomy of the various regions is examined. Beginning with the mouth, the gross and histological structure of each region of the tract is considered. In addition to the mouth, these regions include the esophagus, stomach, small intestine, and large intestine. The major accessory glands, salivary, pancreas, and liver, are also discussed. Having completed a discussion of the structure, the function or physiology of the system is considered. This includes the chemical digestion of the three major classes of organic compounds, carbohydrates, lipids, and proteins, as well as water and electrolyte absorption. Finally, the various vitamins are discussed.

CHAPTER OUTLINE

A. Introduction
　1. Histological organization
　　a. Basic structural patterns
B. The functional anatomy of the digestive system
　1. The oral cavity
　　a. Gross anatomy
　　b. Histological organization
　　c. Special regional anatomy
　　　(1) The tongue
　　　(2) Salivary glands
　　　(3) Teeth
　　　(4) Mastication
　2. The pharynx
　3. The esophagus
　　a. Peristalsis and swallowing
　　　(1) The swallowing reflex
　4. The stomach
　　a. Gross anatomy
　　b. Histological organization
　　c. The regulation of gastric function
　5. The liver, pancreas, and small intestine
　　a. The liver
　　　(1) Gross anatomy
　　　(2) Hepatic circulation
　　　(3) Histological organization
　　b. The pancreas
　　　(1) Gross anatomy
　　　(2) Histological organization
　　　(3) Pancreatic physiology
　　c. The small intestine
　　　(1) Gross anatomy
　　　(2) Histological organization
　　　(3) Intestinal physiology
　　　　(a) Intestinal movements
　6. The large intestine
　　a. Gross anatomy
　　b. Histological organization
　　c. Colonic physiology
　7. The rectum and anus
　　a. Defecation
C. The physiology of digestion
　1. The processing and absorption of nutrients
　　a. The digestion of carbohydrates
　　b. The digestion of lipids

        c. The digestion of proteins.
   2. Water and electrolyte absorption
   3. Vitamins
D. Clinical patterns
   1. Disorder of accessory organs
      a. Gallbladder problems
      b. Liver disorders and clinical tests
         (1) Hepatitis
         (2) Cirrhosis
         (3) Analysis of liver structure and function
   2. Disorders of the digestive tract
      a. Achalasia and esophagitis
      b. Vomiting and intestinal evacuation
      c. Diarrhea and constipation
      d. Diverticulitis and colitis
      e. Intestinal obstruction

LEARNING ACTIVITIES

Complete each of the following items by supplying the appropriate word or phrase.

1. List the six steps involved in the digestive process.

   a. _____
   b. _____
   c. _____
   d. _____
   e. _____
   f. _____

2. The six layers that make up the wall of the digestive tract are, from the inside to the outside, the _____, _____, _____, _____, _____, and the _____.

3. The epithelium of the mouth, esophagus, and anus is a stratified _____ epithelium, while the remainder of the tract is lined by a simple _____ epithelium.

4. The epithelium lies over a loose layer of connective tissue termed the _____ , which, combined with the epithelium, forms the _____ or lining of the digestive tract.

5. The outer boundary of the mucosa is marked by a thin layer of smooth muscle and elastic fibers known as the muscularis _____.

6. The _____ contains large blood vessels and lymphatics, exocrine glands in some regions, and as an outer boundary, a nervous plexus for regulating digestive tract activity.

7. The muscularis externa layer has a layer of circular smooth muscle and a layer of longitudinal smooth muscle. Between these layers is found the _____ plexus which regulates the activity of the muscle layers.

8. The outer layer of dense connective tissue which covers those portions of the tract which lie outside of the peritoneal cavity is termed the _____ while the equivalent layer for those portions which lie inside of the peritoneal cavity is known as the _____.

9. The serosa is also known as the _____ peritoneum.

10. The sheets of serous membrane that connect the visceral and parietal peritoneal membranes together forming suspensions for the digestive tract within the peritoneal cavity are the _____.

11. The four functions of the oral cavity are _____, _____, _____, and _____.

12. Match the terms concerning the oral cavity which are listed below with the appropriate descriptive statement which follows.

    a. cheeks b. labia c. vestibule d. lingual frenulum e. uvula f. sublingual glands g. parotid glands h. submandibular glands

    _____ form the lateral walls of the oral or buccal cavity

    _____ glands that lie in the floor of the mouth

    _____ another name for the lips

    _____ the space between the cheek or lips and the teeth

    _____ lie on the floor of the mouth beneath the epithelium

    _____ found beneath the zygomatic arches

    _____ a fold of mucus membrane that connects the anterior portion of the tongue to the underlying epithelium

    _____ fleshy posterior extension of the soft palate

13. The three major functions of the tongue are _____, _____, and _____

14. The tongue is divided into an anterior body and a posterior _____.

15. The _____ tonsils are found on the root of the tongue.

16. The salivary glands produce _____ liters of saliva per day.

17. One of the principal functions of saliva is to _____ the contents of the mouth.

18. Saliva dissolves chemicals that stimulate the _____.

19. Salivary secretions also keep the _____ population in the mouth under control.

20. The enzyme found in saliva, _____, breaks down starch.

21. The parotid glands produce a saliva that is rich in _____ and much thicker than that produced by the other glands.

22. The sublingual and submandibular glands produce a saliva that is rich in _____ which is used for lubrication.

23. Nervous stimulation from the _____ division of the ANS increases saliva production.

24. Any object placed in the mouth can increase saliva secretion by triggering a _____ reflex.

25. The salivary nuclei are also influenced by the _____ centers and therefore thinking about food can increase saliva flow.

26. In the space provided below, write the term that best fits the statement that follows.

    _____ the process of breaking up food

    _____ makes up the bulk of each tooth

    _____ the central tooth cavity that receives nerves and blood

    _____ canal that communicates with the pulp cavity

    _____ the tooth sits in this bony socket

_____ connects the root to the bone of the socket

_____ substance that covers the root

_____ part of the tooth covered by enamel

_____ blade shaped teeth at the front of the mouth

_____ conical teeth with pointed tips

_____ teeth with three roots

_____ there are usually twenty of these first teeth

_____ last teeth to appear in the adult

_____ replace the deciduous molars in the adult

_____ sticky deposits on teeth produced by bacterial actions

_____ the most common cause of tooth loss

27. The food mass that is formed by the processes of salivation and mastication is known as a _____.

28. Label figure 24.1, the mouth, using the terms listed below. Color the various structures. Refer to figures 24.3 and 24.5 in your textbook for assistance.

cheeks   palatine tonsil   lips (labia)   incisors   canine

premolars   molars   uvula   hard palate   oropharynx

29. Label figure 24.2, the tongue, using the terms listed below. Color each of the structures. Refer to figure 24.3 in your textbook for assistance.

epiglottis   palatine tonsil   fungiform papillae   filiform papillae

circumvallate papillae   lingual tonsil

30. The muscles of the pharynx which provide the impetus for bolus movement are the pharyngeal _____.

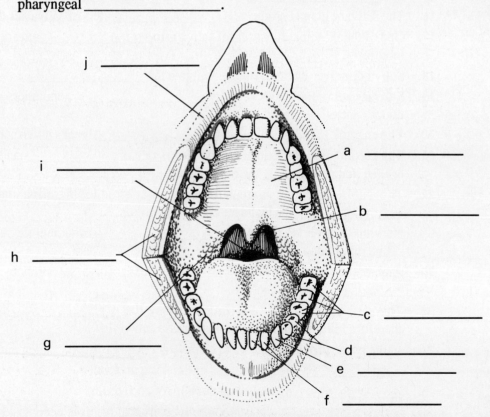

Figure 24.1

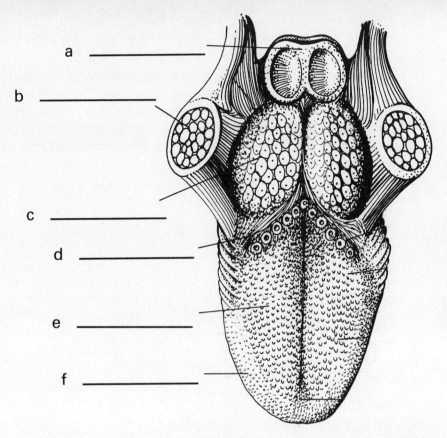

a _____

b _____

c _____

d _____

e _____

f _____

Figure 24.2

31. The upper third of the esophagus contains _____ muscle fibers, the middle third contains smooth and skeletal fibers, and the bottom third contains _____ muscle fibers only.

32. The wave of muscular contraction that pushes material through the digestive tract is known as _____.

33. The three phases of the swallowing process are _____, _____, and _____.

34. The buccal phase begins with the compression of the bolus against the _____.

35. During the pharyngeal phase, the larynx is elevated, closing off the _____ with the epiglottis.

36. The esophageal phase begins with the opening of the upper _____ sphincter.

37. The peristaltic wave triggers the opening of the lower _____ sphincter.

38. Once the bolus has passed to the laryngopharynx, all aspects of swallowing become _____.

39. Secondary peristaltic waves can be initiated in the esophagus and the rest of the tract independent of the CNS by _____ reflexes which are centered in the nervous plexuses found in the walls of the tract.

40. The stomach has three major functions. They are

    a. _____

    b. _____

    c. _____

41. The secretions of the stomach are known as _____ juices and the liquid material produces by stomach activity on ingested foods is termed _____.

42. The stomach is a J shaped organ with the inside of the "J" represented by the __a__ curvature and the outside by the __b__ curvature. The esophagus attaches to the medial surface and the bulge which is superior to the esophageal junction is the __c__. The __d__ lies between the fundus and the curve of the J. The curve of the J is the __e__. The flow of material between the stomach and the small intestine is regulated by the __f__ sphincter.

    a. _____
    b. _____
    c. _____
    d. _____
    e. _____
    f. _____

43. The folds of the empty stomach are termed _____ and they permit the stomach to distend as it fills.

44. The mesentery that attaches to the greater curvature of the stomach is the greater _____.

45. The lesser _____ forms a mesentery connection between the stomach and liver.

46. The neck of the gastric pits are lined by _____ cells.

47. Each gastric pit communicates with a number of _____ glands.

48. The gastric glands have two types of secretory cells, _____ cells secrete HCl and intrinsic factor, while the _____ cells secrete pepsinogen.

49. It is the acid in the stomach that converts pepsinogen into the active form, _____.

50. Pepsin breaks down _____.

51. The hormone _____ stimulates the secretion of gastric juice.

52. The three phases of gastric regulation are the _____, the _____, and the _____ phase.

53. In the _____ phase the sight or thought of food can initiate the flow of gastric juice.

54. In the _____ phase, the arrival of food to the stomach stimulates both stretch and chemoreceptors.

55. During the gastric phase, the _____ externa becomes stimulated, and gastric contractions begin.

56. Most of the regulatory controls of the _____ are inhibitory, and therefore regulate the rate of stomach emptying.

57. Label figure 24.3 using the terms listed below. Color each structure. Refer to figure 24.9 in your textbook for assistance.

duodenum	pyloric sphincter	longitudinal muscle layer
diaphragm	esophagus	greater omentum
greater curvature	oblique muscle layer	circular muscle layer
serosa	fundus	rugae

58. The _____ is the largest visceral organ.

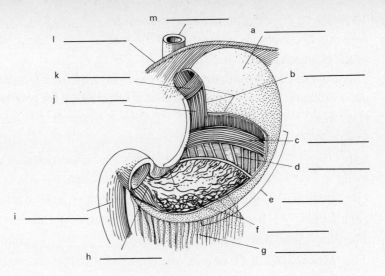

Figure 24.3

59. Over _____ functions have been assigned to the liver. List the three major categories into which these functions can be placed.

    a. _____

    b. _____

    c. _____

60. Liver bile contains water, electrolytes, _____, and lipids.

61. Bacterial action in the digestive tract converts bilirubin into _____.

62. The lipids in bile are collectively called _____.

63. The secretion and reabsorption of the bile salts is referred to as the _____ circulation of bile.

64. Match the term below with the appropriate descriptive statements which follow.

    a. capsule   b. falciform ligament   c. gallbladder   d. cystic duct

    e. hepatic duct   f. common bile duct

    _____ conveys bile into the duodenum

    _____ fibrous covering of the liver that represents the visceral peritoneum

    _____ sac located in the quadrate lobe that stores bile

    _____ drains the gall bladder

    _____ duct that drains bile from the liver lobules

    _____ marks the division between the left and right lobes of the liver

65. The hepatic artery is a branch of the _____.

66. The basic functional unit of the liver is the _____.

67. Each lobule is organized so that the hepatocytes are arranged in plates around a central _____.

68. _____ lie between the plates and empty into the central vein.

69. Blood reaches the sinusoids via branches of the _____ vein and hepatic artery.

70. The hepatocytes absorb and secrete materials into the _____, and _____ is secreted into small channels or canaliculi.

71. The central veins merge to form the _____ veins which enter the inferior vena cava.

72. The release of bile is stimulated by the hormone _____.

73. Bile salts facilitate the absorption of lipids by _____ the large lipid droplets into smaller, absorbable ones.

74. Label figure 24.4, gross anatomy of the liver, using the terms listed below. Color each structure. Refer to figure 24.13 in your textbook for assistance.

right lobe   left lobe   falciform ligament   inferior vena cava
gallbladder   cystic duct   quadrate lobe   caudate lobe

75. Label figure 24.5, liver lobule, using the terms listed below. Color each structure. Refer to figure 24.14 in your textbook for assistance.

bile canaliculi   sinusoid   branch of hepatic artery
central vein   hepatocyte plates branch of portal vein

76. The pancreas secretes both enzymatic and _____ into the small intestine.

77. The pancreatic duct enters the small intestine in association with the _____ duct.

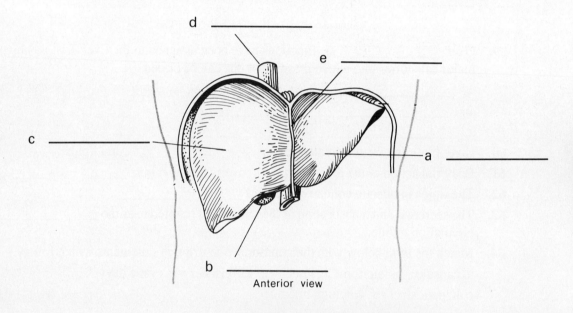

Anterior view

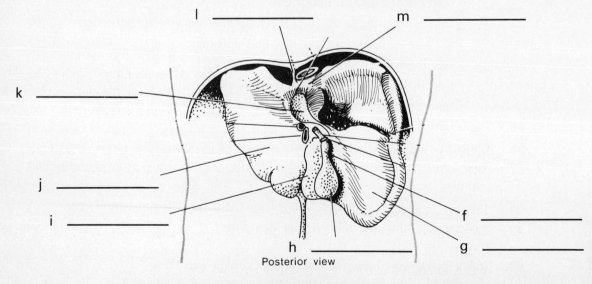

Posterior view

Figure 24.4

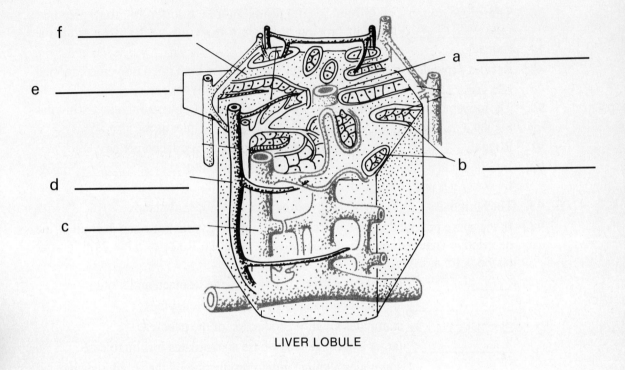

f _____

e _____

d _____

c _____

a _____

b _____

LIVER LOBULE

Figure 24.5

78. The hormone _____ stimulates release of the pancreatic alkaline fluid while the hormone _____ triggers the release of enzymes.

79. _____ account for about 70 percent of pancreatic enzyme production.

80. In order to protect themselves from the powerful proteinases, pancreatic cells secrete them in the form of _____.

81. The small intestine is about _____ meters in length.

82. The small intestine has three segments, the _____, the _____, and the _____.

83. The permanent folds of the intestinal mucosa are the _____.

84. The mucosa of the small intestine is organized into fingerlike projections termed the intestinal _____ which greatly increase the surface area available for absorption.

85. The epithelial cells of each villus have their exposed surfaced covered with _____.

86. In addition of a capillary bed, each villus contains a _____.

87. The intestinal _____ contains cells that produce both enzymes and digestive hormones.

88. There are extensive submucosal glands in the duodenum that produce _____ which protects the lining from the acid chyme from the stomach.

89. Besides peristalsis, regions of the small intestine also undergo a movement termed _____ which mixes and fragments the chyme.

90. The movement of the small intestines is controlled mainly by neural reflexes involving the _____ and _____ plexuses.

91. Roughly _____ liters of intestinal juice are produced per day.

92. Intestinal juice moistens chyme, assists in buffering, and _____ digestive enzymes and products of digestion.

93. The hormone _____ stimulates the duodenal glands.

94. In the space provided below, write the name of the gastric hormone in front of the descriptive statement which it most closely matches. Refer to table 24.5 in your textbook for assistance.

    _____ stimulates gastric secretion by the stomach and motion

    _____ stimulates production of pancreatic enzymes

    _____ stimulates alkaline production of the pancreas

    _____ target is the pancreas where it stimulates insulin release

    _____ stimulates alkaline mucus production by the small intestine

95. The principle functions of the large intestine include

    a _____

    b _____

    c _____

96. The opening between the small intestine and the large intestine is guarded by the _____ valve.

97. The slender finger-like appendage of the cecum is the _____.

98. The five major regions of the colon, in order from the ileo-cecal valve, are the _____, _____, _____, _____, and _____.

99. The crypts of the large intestine are dominated by mucus secreting _____ cells.

100. The internal anal sphincter is composed of _____ and the external anal sphincter is composed of _____ muscle.

101. In addition to normal peristalsis, the colon also exhibits a very rapid and forceful movement known as _____ peristalsis.

102. The elimination of feces from the colon is initiated by the _____ reflex, but also requires voluntary release.

103. The chemical breakdown of complex organic nutrients into absorbable fragments destroys any _____ characteristics they may have and therefore insures that the body will not initiate an immune response against them.

104. All of the major organic molecules, proteins, lipids, nucleic acids, and carbohydrates consist of subunit molecules which are held together by oxygen atoms much in the same fashion that a railroad train is built up of similar coupled cars. During chemical digestion these subunits are freed by adding water across the bond that holds them together, a process known as _____.

105. Carbohydrate digestion begins in the _____.

106. The enzymes of the mouth and pancreas reduce starches to disaccharides and trioses, but it is the enzymes produced by the _____ that breaks these down into monosaccharides which can be absorbed.

107. Complex polysaccharides such as cellulose which cannot be digested pass to the colon where _____ break them down.

108. When colonic bacteria act on undigested organic molecules they produce intestinal gas known as _____.

109. _____ are the most abundant dietary lipid.

110. Within the epithelial cells, the fatty acids are reconverted into triglycerides which are then coated with protein to form _____.

111. Most chylomicrons which are formed are to large to be absorbed into the capillaries and instead are absorbed by the _____ found in each intestinal villus.

112. Of the total 8 - 10 liters of water that enter the digestive tract each day, only about _____ ml gets lost with fecal wastes.

113. The great bulk of water is reabsorbed in the small intestines by means of _____.

114. The active transport of organic nutrients and electrolytes by the intestinal mucosa establishes the osmotic gradient that _____ follows.

115. The overwhelming bulk of all absorption, water, electrolytes, and organic nutrients occurs in the _____.

116. Match the digestive enzyme listed below with the appropriate descriptive statement which follows. Refer to table 24.6 in your textbook for assistance.

a. alpha amylase b. pepsin c. rennin d. trypsin e. chymotrypsin f. carboxypeptidase g. elastase h. lipase i. nuclease j. enterokinase k. sucrase l. peptidase

_____ pancreatic enzyme that hydrolyzes nucleic acids

_____ protein digesting enzyme of the stomach

_____ enzyme found in the the stomach of infants only

_____ secreted by the pancreas as the proenzyme chymotrypsinogen

_____ pancreatic enzyme that breaks down the protein elastin

_____ enzyme found in the small intestine that breaks down sucrose into glucose and fructose

_____ enzyme that is produced by both the mouth and pancreas

_____ pancreatic enzyme which activates other pancreatic proteinases

_____ intestinal enzyme that activates trypsin

_____ enzyme of the small intestines that converts dipeptides and tripeptides into amino acids

_____ pancreatic enzyme that digests proteins and polypepetides into short peptide chains

117. The _____ soluble vitamins function primarily as cofactors in enzymatic reactions.

118. Label figure 24.6, the digestive tract, using the terms listed below. Color each structure. Refer to figure 24.1 in your textbook for assistance.

appendix	cecum	ileum	ascending colon
liver	mouth	rectum	sigmoid colon
parotid gland	pharynx	esophagus	stomach
transverse colon	jejunum	descending colon	duodenum
submandibular and submaxillary glands	pancreas		

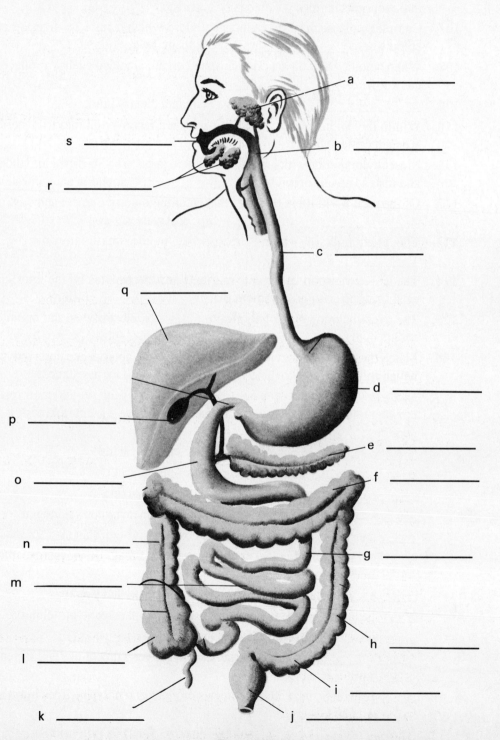

Figure 24.6

119. Match the clinical condition listed below with the appropriate descriptive statement which follows

a. cholecystitis   b. jaundice   c. hepatitis   d. cirrhosis   e. achalasia   f. vomiting
g.emetics   h. dirrhea   i. constipation   j. diverticulitis

_____ condition where pockets form in the mucosa

_____ drugs which promote vomiting.

_____ caused by slow movement of material through the colon which leads to excessive water reabsorption

_____ results from slow peristaltic waves in the esophagus

_____ irritation to the stomach mucosa results in this reflexive emptying

_____ caused by a blockage of the cystic or bile ducts by gallstones

_____ yellowing of the skin and eyes due to backups in the circulation of bilirubin

_____ a viral induced inflammation of the liver

_____ caused by hepatocyte destruction, usually by chronic exposure to alcohol

_____ results when the colonic mucosa is unable to maintain normal levels of absorption

SELF TEST

Circle the correct answer to the following questions.

1. The innermost layer of the digestive tract wall is the
   a. submucosa.   b. muscularis externa.   c. mucosa.   d. adventitia.   e. mesentery.

2. The lining of the mouth consists of a
   a. simple squamous epithelium.   b. pseudostratified columnar epithelium.
   c. cuboidal epithelium.   d. stratified columnar epithelium.   e. stratified squamous epithelium.

3. The basic structure of the tooth is formed from
   a. enamel.   b. bone.   c. dentin.   d. calcium phosphate.   e. none of the above.

4. The adult teeth include
   a. incisors.   b. molars.   c. premolars.   d. canines.   e. more than one of the above is correct.

5. The chief cells of the stomach secrete
   a. hydrochloric acid.   b. intrinsic factor.   c. bicarbonate.   d. pepsin.   e. more than one of the above is correct.

6. During the _____ phase of gastric control, thinking of food will initiate gastric juice flow.
   a. intestinal   b. cephalic   c. gastric   d. esophageal   e. peptic

7. The liver bile contains
   a. enzymes.   b. hormones.   c. bile salts.   d. carbohydrates.   e. more than one of the above is correct.

8. Pancreatic juice contains
   a. chymotrypsin.   b. trypsin.   c. alpha-amylase.   d. elastase.   e. more than one of the above is correct.

9. The bulk of absorption in the small intestine occurs in the
   a. duodenum.   b. jejunum.   c. ileum.   d. pylorus.   e. appendix.

10. The appendix is an appendage of the

    a. transverse colon.   b. sigmoid colon.   c. cecum.   d. descending colon.   e. rectum.

11. Proteins are digested by

    a. sucrase.   b. lipase.   c. nuclease.   d. pepsin.   e. maltase.

12. Lipids are absorbed into the lacteal in the form of

    a. micelles.   b. chylomicrons.   c. microns.   d. phospholipids.   e. lipoproteins.

13. The bulk of water absorption occurs in the

    a. stomach.   b. small intestine.   c. colon.   d. pancreas.   e. esophagus.

14. Enterocrinin stimulates the production and release of

    a. acids and enzymes.   b. acids.   c. enzymes.   d. alkaline mucus.   e. pancreatic enzymes.

15. Chronic exposure of the liver cells to alcohol can cause

    a. gallstones.   b. hepatitis.   c. cirrhosis.   d. enteritis   e. cholera.

CHAPTER 25

Metabolism and Energetics

OVERVIEW

In the previous chapter you learned how the body obtains the nutrients which its cells require. In this chapter you will discover exactly what it does with these nutrients. Basically, cells must obtain energy, usually in the form of ATP, and then use this energy to synthesize the carbohydrates, proteins, lipids, and nucleic acids which they require. You have already seen the details of these processes in chapter three when you examined cell biology. This chapter looks at the broader aspects of these processes, especially the interaction of the various organs and tissues in managing the various metabolic processes.

The chapter begins with a review of the fundamentals of cellular metabolism that were detailed in Chapter Three. Following this review, the metabolic interactions are discussed. These interactions are placed into two broad categories, the absorptive state, which examines the metabolic events which occur during the absorption of nutrients from the intestine, and the postabsorptive state, which examines nutrient processing after absorption into the body is complete. The liver is the central metabolic hub for both states and your need to pay particular attention to its role. The next major topic of consideration is diet and nutrition in which you will learn which nutrients are required and what their dietary source is. The discussion then proceeds to bioenergetics or the amount of energy generated by various nutrients. As the energy content of nutrients is almost always expressed as a measure of the heat which they can release, this topic logically ties into the concept of thermoregulation of body temperature, and the chapter concludes with a discussion of this topic.

As a special note, you should pay particular attention to a clinical comment, nutrition and nutritionists, found on page 975 of your text. Nutritional quackery is one of the growth industries in the United States today. You should insure that you know what constitutes proper nutrition and be able to recognize nutritional snake oil when you see it!

CHAPTER OUTLINE

A. Introduction
B. Metabolic interactions
    1. The absorptive state
        a. The liver
        b. Adipose tissue
        c. Skeletal muscle, neural tissue, and other peripheral tissues
    2. The postabsorptive state
        a. The liver and gluconeogenesis
            (1) The utilization of lipids
                (a) Ketone bodies
            (2) The utilization of amino acids
        b. Adipose tissue
        c. Skeletal muscle
        d. Other peripheral tissues
        e. Neural tissue
    3. Adjustments to starvation
C. Diet and nutrition
    1. The four basic food groups
        a. Nitrogen balance
    2. Minerals, vitamins, and water
        a. Minerals and vitamins
        b. Water
        c. Water and weight loss
    3. Diet and disease
D. Bioenergetics
    1. Thermoregulation
        a. Mechanisms of heat transfer
            (1) Radiation
            (2) Conduction
            (3) Convection
            (4) Evaporation
        b. Mechanisms for increasing heat loss
            (1) Physiological mechanisms
            (2) Behavioral modifications
        c. Mechanisms for promoting heat gain
            (1) Physiological mechanisms
            (2) Behavioral modifications
        d. Thermoregulatory problems in infancy
        e. Thermoregulatory variation among adults
        f. Fevers
E. Clinical patterns
    1. Dietary disorders
    2. Genetic disorders
    3. Storage and transport disorders
    4. Pathogenic disorders

LEARNING ACTIVITIES

Complete each of the following items by supplying the appropriate word or phrase.

1. The energy that cells require is obtained through the breakdown of _____.

2. In order for cells to function, essential nutrients are absorbed by the _____, oxygen is provided by the _____ system, and both are distributed by the _____ system.

3. The _____ system adjusts and coordinates the metabolic activities of the body's tissues, and controls the storage and mobilization of nutrient reserves.

4. The five metabolic components that the body can be divided into are as follows.

   a. _____

   b. _____

   c. _____

   d. _____

   e. _____

5. Following a meal, the absorption of nutrients continues for about _____ hours.

6. Following a meal, blood glucose levels do not rise dramatically because the _____ removes most of it from the blood, storing the excess in the form of glycogen.

7. If excess glucose still remains in the blood after maximum glycogen synthesis, the liver will convert it into _____.

8. The liver can synthesize many of the _____ acids required for protein synthesis and also has the enzymes necessary for the synthesis of _____ used in lipid synthesis.

9. _____ seem to transport cholesterol to the peripheral tissues and _____ transport excess cholesterol back to the liver.

10. During the absorptive state, adipocytes remove _____ and _____ acids from the circulation.

11. Adipocytes can also utilize _____ and _____ acids for lipid synthesis, but they do so only if the levels in the blood are very high.

12. In resting skeletal muscle cells, a portion of the energy demand is met by catabolizing fatty acids and glucose molecules are used to build _____ reserves.

13. During the postabsorptive state, when blood glucose levels begin to decline, the liver begins the breakdown of _____.

14. Once glycogen reserves are depleted, the liver responds to further drops in blood glucose levels by the process called _____ which builds glucose molecules from other carbon fragments.

15. The liver obtains energy for gluconeogenesis by removing and breaking down fatty acids into _____ which then enters the Kreb's cycle.

16. Acetyl CoA is also used to synthesize _____ bodies.

17. Ketone bodies synthesized by the liver are transported to the peripheral tissues where they are reconverted into _____ which is then used for energy.

18. Amino acids that can be deaminated and converted into pyruvic acid are known as the _____ amino acids.

19. The amino acids that can be converted only to acetyl CoA are known as the _____ amino acids.

20. Deamination reactions can only occur in the _____.

21. With declining blood glucose levels, adipocytes begin to release _____ acids and glycerol into the blood stream.

22. In the postabsorptive state, skeletal muscle first utilizes glycogen for energy but later begins to use fatty acids and _____ bodies as energy sources.

23. In the postabsorptive state, neural tissue continues to use _____ as the principal energy source.

24. In extreme starvation, over 90 percent of the body's energy requirement is obtained from the oxidation of _____ bodies.

25. _____ proteins are the last energy source to be mobilized during starvation.

26. A _____ diet contains all of the nutrients required to maintain homeostasis.

27. List the four basic food groups.

   a. _____

   b. _____

   c. _____

   d. _____

28. Those proteins that contain all of the essential amino acids are said to be _____ proteins.

29. Complete proteins are usually found in beef, fish, poultry, eggs, and _____.

30. Compounds that contain nitrogen are termed _____ compounds.

31. Protein sparers include _____ and _____.

32. A person whose intake of nitrogen was equal to the amount excreted would be in _____ balance.

33. A pregnant woman would most likely be in _____ nitrogen balance and consequently would require more nitrogen in her diet than she was excreting.

34. A person who was starving would be in _____ nitrogen balance.

35. Match the mineral listed below with the appropriate descriptive statement which follows. Refer to Table 25.5 in your textbook for assistance.

   a. sodium   b. potassium   c. chloride   d. calcium   e. phosphorus   f. magnesium
   g. iron   h. zinc   i. copper   j. manganese

   _____ cofactor required for hemoglobin synthesis

   _____ major cation in the cytoplasm

   _____ important component of bones, high energy compounds, and nucleic acids

   _____ major cation of the extracellular fluids

   _____ key component of hemoglobin, myoglobin, and the cytochromes

   _____ essential for normal nerve and muscle function, it is also a structural part of bone

   _____ important cofactor for many enzymes

   _____ cofactor for some enzymes

   _____ cofactor for carbonic anhydrase

   _____ major anion of the body

36. Match the vitamin listed below with the appropriate descriptive statement which follows. Refer to Tables 25.6 and 25.7 in your textbook for assistance.

a. A b. D c. E d. K e. thiamine f. riboflavin g. niacin h. pyridoxine
i. folic acid j. cobalamin k. biotin l. pantothenic acid m. ascorbic acid (C)

_____ found in eggs and meat, it functions as a coenzyme for decarboxylation

_____ essential for the synthesis of the clotting factors

_____ an excess of this vitamin leads to hypotension

_____ coenzyme that delivers hydrogen ions

_____ vitamin that is part of NAD

_____ required for normal bone growth

_____ a coenzyme in amino acid and nucleic acid metabolism

_____ deficiency of this vitamin results in pernicious anemia

_____ required for the synthesis of the visual pigments

_____ prevents the breakdown of vitamin A and fatty acids

_____ part of FMN and FAD

_____ vitamin that is part acetyl-CoA

_____ functions as a coenzyme in amino acid and lipid metabolism

37. _____ determines the total amount of energy released when the bonds of an organic molecule are broken.

38. The definition of a _____ is the amount of energy required to raise the temperature of one gram of water one degree centigrade.

39. In nutrition, the Calorie is used to determine energy content and it is equal to _____ standard calories.

40. There are _____ Calories per gram of lipid, _____ per gram of protein, and _____ per gram of carbohydrate.

41. The complete catbolism of glucose by a cell has an efficiency of about _____ which means that of the total energy available in the glucose, _____ percent is lost as heat.

42. The minimum, rising energy expenditures of an awake, alert person is their _____ metabolic rate or BMR.

43. An average person has a BMR of _____ Calories per hour.

44. Heat exchange with the environment involves four basic processes: _____, _____, _____, and _____.

45. Warm objects lose heat as _____ radiation.

46. Over _____ of average heat loss is by radiation.

47. Transfer of energy by direct physical contact is known as _____.

48. _____ is the conductive heat loss to the air that overlies the surface of the body.

49. When water changes from a liquid to a vapor it absorbs _____.

50. Evaporative heat loss is by means of _____.

51. Increased atmospheric _____ retards evaporation of perspiration and therefore retards heat loss.

52. The source of most body heat is _____ reactions.

53. The heat generated by the body is absorbed by the body's _____ and is distributed evenly.

54. The _____ area of the hypothalamus is the body's thermostat.

55. The thermostat affects two areas, a _____ center and a _____ center.

56. When the temperature of the preoptic nucleus exceeds the thermostat setpoint, the __a__ center is stimulated. This results in inhibition of the vasomotor center which causes __b__, and warm blood flows to the skin. As the skin increases in temperature, __c__ and __d__ losses increase. The __e__ are stimulated and perspiration flows on to the surface of the skin where an evaporative cooling takes place. The respiratory center becomes stimulated and the __f__ of respiration increases thereby increasing evaporative heat losses from the lungs.

    a. _____
    b. _____
    c. _____
    d. _____
    e. _____
    f. _____

57. When the temperature of the preoptic nucleus is less than the thermostat setpoint, the vasomotor center __a__ blood flow to the skin which reduces __b__, __c__, and __d__, losses of heat. Special arterial venous shunts trap body heat close to the body core by means of __e__ heat exchange traps. __f__ thermogenesis increases heat production by increasing muscle tone. __g__ thermogenesis increases heat production by an increase in overall metabolic rate via release of the hormone __h__.

    a. _____
    b. _____
    c. _____
    d. _____
    e. _____
    f. _____
    g. _____
    h. _____

58. In both heat and cold stress, _____ modifications such as seeking appropriate environments aid in body temperature regulation.

59. _____ is the making of physiological adjustments to a particular environment over time.

60. Infants do not shiver, but they can generate extra heat by metabolizing _____ fat which produces pure heat and not ATP.

61. Obese individuals have difficulty losing body heat because they have a relatively undersized _____ area.

62. A _____ is the maintenance of a body temperature greater than 37.2 degrees c.

63. Circulating proteins called _____ can reset the hypothalamic thermostat and result in fever.

64. _____ is an endogenous pyrogen.

65. When the thermostat setting is raised, the _____ center is activated and temperature increasing events such as shivering begin.

66. Following the resetting of the thermostat to normal, the _____ phase occurs and the heat-loss center is activated.

67. Fevers seem to be _____ in fighting various infectious diseases.

68. Aspirin brings down fever and is an example of an _____ drug.

69. Match the clinical condition listed below with the appropriate descriptive statement which follows.

a. marasmus   b. kwashiorkor   c. regulatory obesity   d. metabolic obesity
e. PKU   f. gout

_____ obesity that is a result of overconsumption of food

_____ condition which is due to hyperuricemia

_____ obesity that is due to a failure in metabolic regulation

_____ inadequate protein in the diets of infants

_____ protein deficiency in children

_____ genetic disorder in which the amino acid phenylalanine is not properly metabolized

SELF TEST

Circle the correct answer to the questions below.

1. The focal point for all metabolic regulation and control is the
   a. small intestine.   b. liver.   c. kidney.   d. endocrine system.   e. nervous system.

2. Neural tissue is dependent upon _____ for energy.
   a. acetyl CoA   b. beta oxidation   c. anaerobic glycolysis   d. aerobic glycolysis
   e. lactic acid.

3. The absorptive state lasts for about _____ hours following a meal.
   a. 2   b. 3   c. 4   d. 5   e. 12

4. In the postabsorptive state, the liver maintains blood glucose levels initially by breaking down
   a. lipids.   b. starch.   c. glycogen.   d. cellulose.   e. protein.

5. During lipid catabolism, acetyl CoA can form _____ which then diffuse to the peripheral tissues where they are again converted into acetyl CoA.
   a. fatty acids   b. triglycerides   c. glycerol   d. ketone bodies   e. glucose

6. Glucogenic amino acids can be converted into _____ by the liver.
   a. fatty acids   b. lactic acid   c. pyruvic acid   d. glycerol   e. ketone bodies

7. A key feature in the conversion of amino acids into glucose which is carried out by the liver is
   a. decarboxylation.   b. deamination.   c. removal of sulfur.   d. beta oxidation.
   e. none of the above.

8. The last molecules to be catabolized during starvation will be the
   a. sugars.   b. glycogens.   c. fatty acids.   d. triglycerides.   e. structural proteins.

9. Which of the following is not an N compound?
   a. amino acids   b. purines   c. pyrimidines   d. creatine   e. glycogen

10. _____ is a mineral which is required as a cofactor in hemoglobin synthesis.
    a. Manganese   b. Iron   c. Zinc   d. Copper   e. Potassium

11. The vitamin which forms a part of both FMN and FAD is
    a. thiamine.   b. riboflavin.   c. niacin.   d. pyridoxine.   e. folic acid.

12. _____ forms a part of acetyl CoA.

a. Ascorbic acid   b. Biotin   c. Folic acid   d. Pantothenic acid   e. Pyridoxine

13. The greatest amount of energy per gram is contained in

a. carbohydrates.   b. lipids.   c. proteins.   d. nucleic acids.   e. glucose.

14. Which of the following is a mechanism by which the body loses heat?

a. conduction   b. convection   c. radiation   d. evaporation   e. more than one of the above is correct

15. Protein deficiency in children is known as

a. pyrexia.   b. marasmus.   c. Kwashiorkor.   d. PKU.   e. ketonemia.

CHAPTER 26
The Urinary System

OVERVIEW

The urinary system is the last of the systems that are directly responsible for maintaining homeostasis of the body fluids. You have already studied the contributions of the circulatory, respiratory, and digestive systems to the maintenance of homeostasis. The urinary system contributes by ridding the body fluids of metabolic wastes and by adjusting the levels of most other substances found in the fluids, including water, electrolytes, and organic nutrients. The urinary system is the major fluid balancing system in the body and contributes more to fluid constancy than any other system. The truth of this last statement can be demonstrated most graphically when failure of the urinary system occurs. Within days of such a failure, there will be a complete loss of homeostasis and death will ensue, even with every other system working perfectly. It maintains the fluid constancy by determining what levels of what substances are optimal, and maintaining these levels either by excretion of excess amounts, or conservation if the quantities are below optimum levels.

This chapter begins by reviewing all of the organs involved in excretion which is followed by a brief description of the gross anatomy of the organs of the urinary system. A detailed examination of the microscopic anatomy of the kidney follows. It is important that you master this material because the functioning of the kidney is very closely tied to its microscopic structure, and it will prove difficult to understand renal physiology without a good understanding of the kidney's histology. The chapter now proceeds to a consideration of renal physiology. The three major functions of the kidney, filtration, reabsorption, and secretion are detailed. This is followed by a thorough discussion of osmosis and how the kidney controls the amount of water that is lost with the urine. As you will see, the ability to regulate the volume of water in the urine is critical to the maintenance of proper water levels in the body. Finally the regulation of kidney function is discussed.

The operation of the kidney is one of the most elegant as well as important physiological areas in the body. Unfortunately it is somewhat complex, and is an area of anatomy and physiology that students often have difficulty with. Mastery of kidney functioning is very important because it is the key to the understanding of fluid, electrolyte, and acid-base balance, topics that will be discussed in the next chapter.

CHAPTER OUTLINE

A. Introduction
B. Gross anatomy
    1. The kidneys
    2. The ureters and bladder
C. Histological organization
    1. The kidney
        a. The nephron
            (1) The renal corpuscle
            (2) The proximal convoluted tubule
            (3) The loop of Henle
            (4) The distal convoluted tubule
        b. The collecting ducts
        c. The blood supply to the nephron
    2. The renal pelvis, ureters, and urinary bladder
    3. The urethra
D. Renal physiology
    1. Filtration
    2. Active transport
        a. Tubular reabsorption and the proximal convoluted tubule
        b. Tubular secretion and the distal convoluted tubule
    3. Osmosis
        a. The formation and maintenance of the osmotic gradient
        b. The production of hypertonic urine
    4. The regulation of kidney function
        a. The control of glomerular filtration
            (1) Autoregulation
            (2) Hormonal regulation
            (3) Autonomic regulation
        b. The control of tubular transport and osmosis
    5. Urine storage and release
E. Clinical patterns
    1. Problems affecting the kidneys
        a. Trouble with the filtration mechanism
        b. Trouble with other tubular functions
    2. Problems with urinary distribution and excretion
    3. Renal failure

LEARNING ACTIVITIES

Complete each of the following items by supplying the appropriate word or phrase.

   1. The urinary system consists of the _____, _____, _____, and _____.

2. In addition to the urinary system the _____, _____, and _____ systems also have excretory roles.

3. The functional organs of the urinary system are the _____.

4. Label figure 26.1, the urinary system, using the terms listed below. Color each organ. Refer to figure 26.1 in your textbook for assistance.

    inferior vena cava   renal vein   ureter   aorta
    urinary bladder   left kidney   renal artery

5. The indentation of the kidney where the vessels enter and exit is the _____.

6. The kidneys lie outside of the periotoneal cavity in a _____ position.

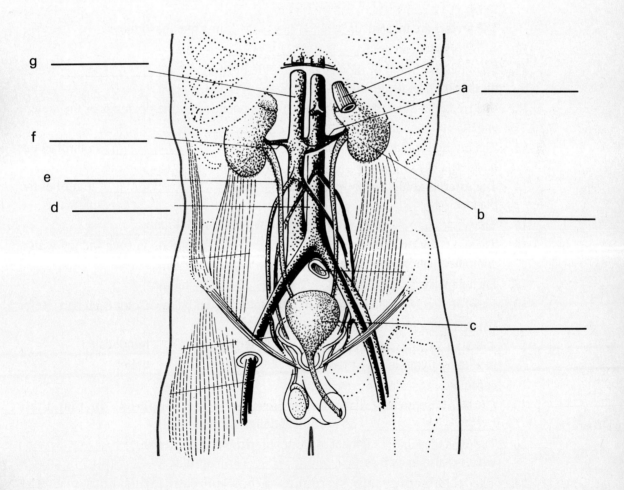

Figure 26.1

7. Match the terms concerning kidney structure which are listed below with the appropriate descriptive statements which follow.

a. cortex b. medulla c. pyramids d. papillae e. renal columns f. interlobar arteries g. arcuate arteries h. interlobular arteries i. renal pelvisim j. major calyces k. minor calyces

_____ enlarged chamber that the ureters open into

_____ each of these small chambers surrounds a papilla

_____ vessels derived from the interlobar arteries

_____ inner layer of the kidney

_____ cortical extensions that dip towards the renal sinus between the pyramids

_____ conical structures which make up the medulla

_____ derived directly from the renal artery

_____ branches of the renal pelvis

_____ outer layer of the kidney

_____ arcuate arteries give rise to these vessels

_____ structures which are the tips of the pyramids

8. The veins found in the kidneys parallel the _____ and are named the same.

9. The ureters are located _____ just like the kidneys.

10. A full bladder can hold _____ of urine.

11. The mucosal lining of the bladder is folded to form _____ which disappear when the bladder is full.

12. The triangular area of the bladder which is bounded by the openings of the ureters and the entrance to the urethra is the _____.

13. The internal sphincter of the urethra provides _____ control of the bladder.

14. The external sphincter of the urethra provides _____ control of the bladder.

15. The _____ is the functional unit of the kidney.

16. There are approximately _____ nephrons in each kidney with a combined length of 85 miles.

17. Each nephron empties into a _____ tubule.

18. Label figure 26.2, the kidney, using the terms listed below. Color each part. Refer to figure 26.3 in your textbook for assistance.

major calyx   cortex   renal column   minor calyx
medullary pyramid   renal pelvis   papilla   ureter
capsule

19. Cortical nephrons differ from juxtamedullary nephrons in that their _____ do not descend into the medulla.

20. The expanded initial segment of the nephron tubule is the renal _____ which is also known as _____ capsule.

21. The renal corpuscle surrounds a ball of capillaries known as the _____.

22. There are three barriers between blood in the glomerulus and the inside of the nephron tubule. They are, in order, the capillary _____, the _____, and the _____ epithelium.

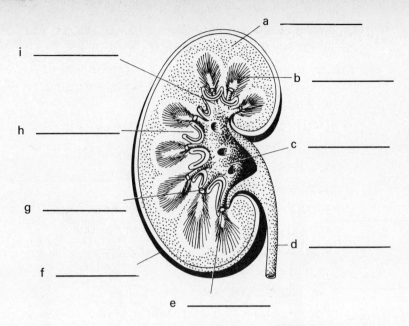

Figure 26.2

23. The capillary endothelium of the glomerulus is the _____ type with pores so large that even proteins will pass across, but not blood cells.

24. The _____ permits the movement of smaller proteins but blocks the larger ones.

25. The cells of the glomerular epithelium are termed _____ and have cellular processes known as _____ which enwrap the outer surface of the basement membrane.

26. It is the _____ pores between the podocytes that block the smaller proteins.

27. Label figure 26.3, the nephron, using the terms listed below. Color each part. Refer to figures 26.5 and 26.6 in your textbook for assistance.

    loop of Henle   afferent arteriole   efferent arteriole
    glomerulus   renal corpuscle   proximal convoluted tubule
    distal convoluted tubule   peritubular capillary bed
    collecting tubule

28. Write the parts of the nephron, in proper order, in the spaces provided below.
    renal corpuscle - _____ - _____ - _____ - collecting tubule

29. Match the parts of the nephron listed below with the appropriate descriptive statements which follow. Refer to Table 26.1 in your textbook for assistance.

    a. renal corpuscle   b. proximal convoluted tubule   c. descending limb   d. ascending limb   e. distal convoluted tubule   f. collecting tubule   g. collecting duct   h. papillary duct

    _____ portion of the loop of Henle that reabsorbs water from the filtrate

    _____ conducts urine to the minor calyx

    _____ where the plasma is filtered

    _____ part of the loop of Henle that reabsorbs electrolytes and assists in creating the medullary concentration gradient

    _____ portion that the distal convoluted tubule empties into, it reabsorbs water and sodium ions

Label: loop of Henle   afferent arteriole   efferent arteriole   glomerulus   renal corpuscle   proximal convoluted tubule   distal convoluted tubule   peritubular capillary bed   collecting tubule

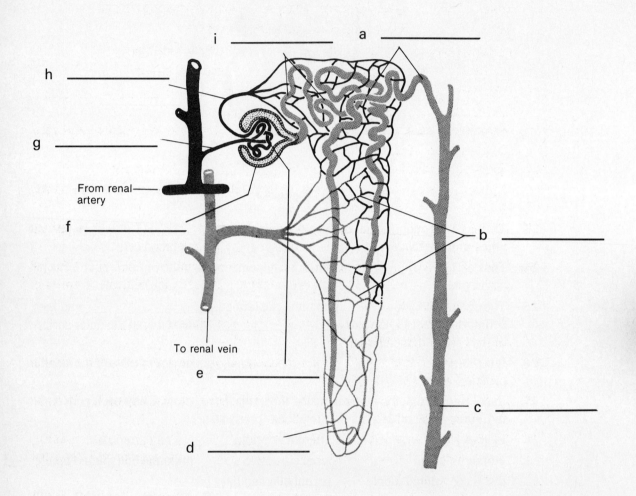

From renal
artery

To renal vein

Figure 26.3

_____ reabsorbs water and secretes hydrogen ions

_____ reabsorption of water and secretion of acids, ammonia, and drugs are all major
        functions

_____ where the bulk of electrolytes, nutrients, vitamins, and water are reabsorbed

30.   The _____ ducts drain groups of collecting tubules.

31.   Collecting ducts unite to form _____ ducts.

32.   The glomerulus is supplied by an _____ arteriole and is drained by
      an _____ arteriole that in turn sends blood to the _____
      capillary bed.

33.   The renal pelvis, ureter, and bladder are lined by a _____ epithelium.

34. The three types of epithelia found lining the urethra are _____,
_____, and _____.

35. The _____ muscle of the bladder is composed of an inner longitudinal layer, a middle circular layer, and an outer longitudinal layer.

36. Label figure 26.4, the ureter, bladder, and urethra, using the terms listed below. Color each part. Refer to figures 26.4 and 26.9 in your textbook for assistance.

fibrous coat   outer longitudinal muscle   transitional epithelium
ureters   inner longitudinal muscle   trigone
circular muscle   lumen   urethral openings
urethra   external sphincter   internal sphincter

37. The three processes involved in urine formation are _____, _____, and _____.

38. Filtration pressure = glomerular blood pressure – (capsular hydrostatic pressure + _____).

39. The average filtration pressure is about _____ mm Hg.

40. The actual amount of filtration that occurs at filtration pressure is known as the _____ filtration rate.

41. On the average, _____ ml of plasma are filtered into the nephrons each minute. This equals _____ liters per day.

42. Even though a large amount of initial urine is formed each day, only about 1 liter of final urine is excreted. This is because 99 percent of the initial urine (filtrate) is _____.

43. The cells of the proximal convoluted tubules reabsorb about _____ percent of the filtrate generated by the glomerulus.

44. The concentration at which the transport enzymes for a substance saturate is known as the _____.

45. Once the transport maximum for a substance has been reached, any further increase in tubular concentration will be lost to the _____.

46. Transport maximum determines the renal _____ for a substance, or the plasma concentration at which a substance will start appearing in the final urine.

47. The proximal convoluted tubules actively reabsorb organic nutrients and _____.

48. The substances which are reabsorbed into the interstitial fluid diffuse into the _____ capillary bed.

49. The active reabsorption of substances by the proximal convoluted tubule creates an osmotic gradient that _____ passively follows.

50. Secretion occurs mainly in the _____ tubule.

51. In secretion, materials are _____ transported into the tubule from the interstitial fluid.

52. Potassium is secreted into the tubules by exchanging it for _____ which is inside of the tubules.

53. Deamination occurs in the cells of the DCT when the blood pH begins to fall. The deamination process produces _____ which combines with hydrogen ions to form _____ which is then excreted.

54. A solution that contains one mole of particles would by a _____ milliosmolar solution.

Label: fibrous coat   outer longitudinal muscle   transitional epithelium   ureters   inner
         longitudinal muscle   trigone   circular muscle   lumen   urethral openings   urethra
         external sphincter   internal sphincter

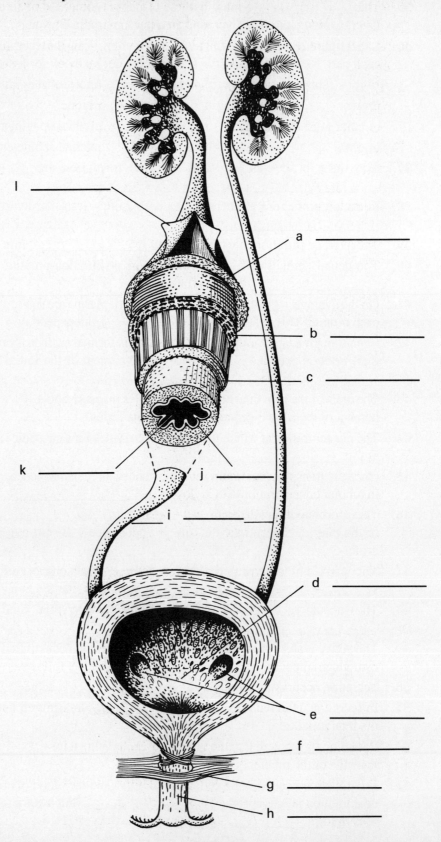

l

a

b

c

k

j

i

d

e

f

g

h

Figure 26.4

55. Water will move across a selectively permeable membrane from a solution of lower _____ to one of higher _____.

56. Assume that solution X is a standard reference in terms of its osmolarity. Solution Y is equal in osmolarity and it would therefore be classed as an _____ solution. Solution Z has a greater osmolarity and it would be a _____ solution. Finally, solution T has a lower osmolarity: it would be a _____ solution.

57. As one moves deeper into the medulla, the osmotic concentration increases and forms an osmotic _____.

58. The osmotic gradient of the medulla is established by the loop of Henle, the vasa recta, and the _____ tubules.

59. The medullary concentration gradient begins with the ascending limb which is __a__ to water and actively transports sodium chloride into the interstitial fluid increasing its concentration. The descending limb is __b__ to water and it follows the concentration gradient established by the active transport from the ascending limb. This removal of water __c__ the urine with solutes which arethen pumped out the ascending limb, further enhancing the concentration gradient. This is an example of __d__ feedback. The concentration gradient is further enhanced by__e__ which diffuses out from the tip of the collecting tubules after it has been concentrated by __f__ withdrawal from the tubules.

   a. _____
   b. _____
   c. _____
   d. _____
   e. _____
   f. _____

60. The _____ contributes to the concentration gradient by removing the water and solutes which would otherwise alter it.

61. Urine leaving the proximal convoluted tubule is greatly reduced in volume from that entering, but is still _____ to the plasma.

62. Through the loop of Henle the tonicity of the urine matches that of the interstitial fluid, reaching a maximum of about _____ milliosmoles at the tip of the loop.

63. As urine moves through the ascending limb it again becomes dilute due to the removal of _____.

64. By the time urine reaches the distal convoluted tubule is has an osmolarity of about _____ mOsm.

65. The _____ tubules pass through the osmotic gradient created by the loop of Henle.

66. If the collecting tubules are permeable to water then it will move out into the osmotic gradients and the urine will become _____, but if the tubules are not permeable to water then all of the water will remain and the urine will be dilute and high in volume.

67. The permeability of the collecting tubules is controlled by _____.

68. If ADH concentrations are high then the collecting tubules are _____ and water is reabsorbed producing a hypertonic urine.

69. If ADH concentrations are low then the collecting tubules are relatively _____ and water will not be reabsorbed, producing a hypotonic urine.

70. The yellow color of urine is due to _____ which is generated by intestinal bacteria and absorbed by the colon.

71. _____ is excessive urine production, _____ is inadequate urine production, and _____ is no urine production.

72. _____ is when there are excessive metabolic wastes in the urine.

73. _____ tests are used to estimate GFR.

74. A reduction in blood pressure would cause the afferent arteriole to the glomerulus to _____ and the efferent arteriole to _____.

75. Sustained low blood pressure will cause a release of renin which then converts angiotensinogen into the active angiotensin II. The overall effects of angiotensin II are as follows.

    a. _____

    b. _____

    c. _____

76. Sympathetic stimulation causes _____ of the afferent arteriole and a _____ in GFR.

77. Urine reaches the bladder by _____ contractions of the ureters.

78. In the bladder, stretch receptors initiate the _____ reflex causing bladder contraction.

79. Match the clinical condition listed below with the appropriate descriptive statement which follows.

    a. polycystic kidney disease   b. nephritis   c. pyelonephritis   d. glomerulonephritis
    e. nephrosis   f. tubular necrosis   g. renal failure   h. hemodialysis

    _____ glomeruli are damaged but still functional

    _____ usually due to inflammation caused by a bacterial infection of the renal pelvis

    _____ tubular functions are affected by this condition which may be due to ischemia, heavy metal poisoning, drugs, and solvents

    _____ genetic condition in which swellings develop along the length of the tubules

    _____ general term for inflammation of kidney tissue

    _____ general loss of all kidney function

    _____ treatment for total renal failure

    _____ This is an inflammation of the cortex which effects the filtering mechanism.

SELF TEST

Circle the correct answer to the questions below.

1. Which of the following plays a role in excretion?
   a. urinary system   b. digestive system   c. respiratory system   d. integumentary system   e. all of the above

2. The organs which convey urine from the kidneys to the bladder are the
   a. urethras.   b. ureters.   c. major calyces.   d. minor calyces.   e. renal pelvis.

3. The tips of the medullary pyramids are the

  a. minor calyces. b. papillae. c. renal columns. d. renal lobes. e. renal sinuses.

4. The arcuate arteries give rise to the

  a. interlobar arteries. b. renal arteries. c. afferent arterioles. d. efferent arterioles. e. interlobular arteries.

5. Each renal pyramid opens into a

  a. renal sinus. b. major calyx. c. renal pelvis. d. minor calyx. e. ureter.

6. Filtration occurs across the

  a. glomerulus. b. proximal convoluted tubule. c. distal convoluted tubule.
  d. loop of Henle. e. collecting tubule.

7. The bulk of reabsorption occurs in the

  a. renal corpuscle. b. proximal convoluted tubule. c. loop of Henle d. distal convoluted tubule. e. collecting tubule.

8. Aldosterone primarily affects the

  a. glomerulus. b. proximal convoluted tubule. c. distal convoluted tubule.
  d. loop of Henle. e. renal corpuscle.

9. The ascending limb of the loop of Henle is

  a. permeable to water. b. permeable to electrolytes. c. the site of active transport of sodium and chloride. d. of no importance in establishing the osmotic gradient of the medulla. e. none of the above.

10. Which of the following plays a role in establishing the osmotic gradient through which the collecting tubules must pass?

  a. vasa recta b. loop of Henle c. urea d. sodium and chloride e. more than one of the above is correct

11. If ADH levels in the blood are high, then the collecting tubules will be

  a. permeable to water. b. impermeable to water. c. permeable to sodium. d. permeable to urea. e. impermeable to protein.

12. If ADH levels are low, then the urine produced will be _____ with respect to the plasma.

  a. hypertonic b. isotonic c. hypotonic

13. Increasing the diameter of the afferent arteriole while decreasing the diameter of the efferent arteriole will.

  a. decrease the glomerular pressure. b. decrease the GFR. c. decrease the volume of blood in the glomerulus. d. increase the GFR. e. none of the above

14. Which of the following would decrease GFR?

  a. aldosterone release. b. ADH release. c. erythropoietin release. d. dilation of the afferent arteriole. e. increased sympathetic activity.

15. Inflammation which occurs within the cortex and affects the filtration activities is known as

  a. pyelonephritis. b. nephrosis. c. tubular necrosis. d. glomerulonephritis.
  e. aminoaciduria.

CHAPTER 27
Fluid, Electrolyte, and Acid-Base Balance

OVERVIEW

The maintenance of body fluid homeostasis is largely a maintenance of the proper levels of three substances, water, electrolytes, and hydrogen ions. The regulation of these substances is carried out by systems which you have already examined, specifically the urinary, respiratory, circulatory, and endocrine systems. This chapter represents a continuation of the discussion of the functions of these systems as they relate to body fluid homeostasis.

The chapter begins with a discussion of the various fluid compartments of the body and how they interact with one another. The presentation then moves to a consideration of water and electrolyte balance. The two are closely related because it is usually alteration of electrolyte levels that lead to osmotic shifts of water. Electrolyte balance is principally a discussion of how sodium and potassium levels are maintained. The last topic is acid-base balance. After defining acid, base, and pH, a discussion ensues as to how the body maintains normal pH. The section and chapter conclude with a look at the various disturbances that can occur in acid-base balance.

CHAPTER OUTLINE

A. Introduction
B. Fluid and electrolyte balance
   1. Fluid balance
      a. Disturbances of fluid balance
   2. Electrolyte balance
      a. Sodium ion regulation

        b. Potassium ion regulation
  C. Acid-base balance
      1. Acid, bases, and pH
          a. The maintenance of normal pH
              (1) Buffers and buffer systems
                  (a) The bicarbonate buffer system
                  (b) The phosphate and protein buffer systems
              (2) Pulmonary contributions to pH regulation
              (3) The renal contribution to pH regulation
      2. Disturbances of acid-base balance
          a. Respiratory acidosis
          b. Respiratory alkalosis
          c. Metabolic acidosis
          d. Metabolic alkalosis
  D. Clinical patterns
      1. The treatment of acidosis
      2. The treatment of alkalosis

LEARNING ACTIVITIES

Complete each of the following items by supplying the appropriate word or phrase.

1. The two major fluid compartments of the body are the _____ and
   the _____.

2. The extracellular fluid compartment can be subdivided into the
   _____ and the _____.

3. The fluid compartments are separated from each other by _____ permeable membranes.

4. The principal electrolytes in the ECF are _____ and _____

5. The principal electrolytes in the ICF are _____, _____,
   and _____.

6. Although the composition of the ECF and ICF are quite different their
   _____ are equal.

7. There are no receptors for fluid or electrolyte balance. Adjustment occurs in response
   to changes in _____ or _____.

8. Fluid and electrolyte balance are mediated by three hormones,
   _____, _____, and _____.

9. Match the hormone listed below with the appropriate descriptive statement which
   follows.
   a. ADH   b. aldosterone   c. atrial natriuretic factor
   _____ released in response to osmoreceptors in the hypothalamus
   _____ increases sodium reabsorption from the distal convoluted tubule
   _____ increases water conservation
   _____ released in response to reduced sodium levels
   _____ will block the release of both ADH and aldosterone
   _____ an increase in the volume of blood to the heart would stimulate its secretion
   _____ decreasing blood volume would stimulate this hormone

10. Approximately _____ ml of water are lost each day in urine, feces, and insensible perspiration.

11. The water losses are normally balanced by gains due to _____, _____, and _____.

12. If the ECF loses water and becomes hypertonic to the ICF, water will move from the _____ to the _____ until they are again isotonic with one another.

13. Water gain can be adjusted by changing the rate of fluid consumption, and water loss can be regulated by the levels of _____ in the plasma.

14. A person has been perspiring profusely due to heat stress. You would expect that his or her ADH would be _____.

15. Eighty percent of the osmolarity of the ECF is due to _____.

16. Sodium losses occur primarily in the _____ and the _____.

17. In the distal convoluted tubule, sodium reabsorption is electrically balanced by either _____ or _____ ion ejection.

18. Alterations in the rate of sodium uptake do not alter sodium concentration of the ECF because of the movement of _____ which always accompanies sodium.

19. If sodium reabsorption from the urine increases, the osmotic reabsorption of water also increases, and the ECF volume _____ but the concentration of sodium remains constant.

20. Ingestion of large amounts of salt may temporarily increase sodium levels in the plasma, but the increased osmolarity will trigger the release of _____ which will cause an increase in water conservation and a reduction in osmolarity.

21. When fluid balance mechanisms are normal, alterations in the electrolyte levels will lead to to expansion or contraction of the ECF _____ which will maintain normal sodium levels.

22. Large drops in the ECF __a__ that result in reduced blood pressure cause a release of __b__ by the kidney. This leads to a production of angiotensin II which then stimulates __c__, causes the release of __d__, and initiates the release of __e__ by the adrenals. Aldosterone promotes the reabsorption of __f__ from the nephron tubules. Increased sodium reabsorption is accompanied by increased __g__ reabsorption, and a restoration of the ECF volume.

   a. _____
   b. _____
   c. _____
   d. _____
   e. _____
   f. _____
   g. _____

23. Salt retention always leads to _____ retention. This is why people with high blood pressure are placed on reduced salt diets.

24. The bulk of potassium is found in the _____.

25. Potassium in the ECF is controlled by diet, tubular reabsorption, and tubular _____.

26. Increased potassium levels in the ECF will cause an increase in the secretion of _____ which in turn increases secretion of potassium into the distal convoluted tubule.

27. As potassium secretion involves the exchange and retention of sodium, elevated sodium reabsorption in the distal tubules will result in _____ plasma potassium levels.

28. Alterations of hydrogen ion causes changes in the shape and activation state of _____.

29. A solution is said to be _____ if the hydrogen ions equal the hydroxyl ions. The solution is _____ if there are a greater number of hydrogen ions than hydroxyl ions, and it is _____ if the number of hydroxyl ions exceeds the number of hydrogen ions.

30. _____ are compounds that dissociate in water and release hydrogen ions while _____ are compounds that dissociate and release hydroxyl ions or remove hydrogen ions from solution.

31. Compounds that have no affect on either hydrogen ion or hydroxyl ion concentration when they dissociate are termed _____.

32. A _____ acid or base will dissociate completely while a _____ acid or base will only partially dissociate.

33. Adding a strong acid to a weak base will result in the formation of a weak _____.

34. Adding hydrochloric acid to sodium bicarbonate produces _____ acid.

35. _____ is the measure of the hydrogen ion concentration of a solution.

36. The maintenance of normal body pH is carried out by _____, _____ mechanisms, and _____ mechanisms.

37. _____ are compounds which will absorb or release hydrogen ion thereby maintaining a stable pH.

38. Most threats to body pH come from the _____ side.

39. There are three general classes of acids in the body, _____ acids, _____ acids, and _____ acids.

40. Volatile acids are those which diffuse out of solution. The most important one in the body, and in fact the most important acid in the body of any type, is _____.

41. Carbon dioxide is considered to be an acid because it reacts with water to form _____ acid.

42. Fixed acids are inorganic acids that stay in solution. Two that occur as a result of metabolism are _____ and _____.

43. _____ acids include fatty acids, ketone bodies, and other intermediates of metabolism.

44. The first line of defense against pH changes in the body fluids are the _____.

45. A buffer system consists of a weak acid and its complementary _____.

46. The three major buffering systems of the body are _____, _____, and _____ buffer systems.

47. The most important buffer of the ECF is _____.

48. Protein can function as a buffer because amino acids can either release or accept a _____ ion depending upon pH of the solution they are in.

49. The lungs and kidneys participate in pH regulation primarily through their influence on the _____ buffer system.

50. When respiration increases, __a__ is lost. This results in a decrease in __b__ acid molecules. These molecules are replaced by combining __c__ ion with bicarbonate. Bicarbonate is obtained from __d__ and the hydrogen ion from other acids. As a result, hydrogen ion __e__, and plasma pH increases.

   a. _____

   b. _____

   c. _____

   d. _____

   e. _____

51. The kidneys can modify pH by secreting _____ and reabsorbing _____.

52. Urinary pH must be kept above _____ and consequently most of the hydrogen ion secreted into the urine combines with _____.

53. _____, _____, and _____ are all buffers in the urine which tie up hydrogen ion.

54. Under conditions of _____, the kidney will stop reabsorbing bicarbonate and secreting hydrogen ion.

55. The pH of the extracellular fluid is almost always between _____ and _____.

56. Respiratory acidosis or _____ occurs when the respiratory system cannot remove sufficient quantities of carbon dioxide and carbonic acid begins to accumulate.

57. The usual cause of respiratory acidosis is _____.

58. Hyperventilation can lead to transitory respiratory _____.

59. _____ acidosis can result from any condition that depletes the normal reserves of bicarbonate ions.

60. The most frequent cause of metabolic acidosis is a loss of ability to excrete hydrogen ions by the _____.

61. Generation of large quantities of ketone bodies can lead to a type of metabolic acidosis known as _____.

62. Chronic diarrhea can lead to metabolic acidosis by depleting _____.

63. Metabolic alkalosis occurs when _____ concentrations become elevated.

64. One possible cause of metabolic alkalosis would be continuous _____.

65. Treatment of acidosis is by increasing pulmonary ventilation and by administering _____.

66. Match the clinical condition listed below with the typical causes that follow. Refer to Table 27.1 in your textbook for assistance.

a. respiratory acidosis   b. respiratory alkalosis   c. metabolic acidosis   d. metabolic alkalosis

_____ excessive diarrhea can cause this

_____ most often due to vomiting or use of diuretics

_____ it can result from hyperventilation

_____ cardiac arrest and congestive heart failure cause this

_____ usually results from renal failure

_____ can result from emphysema or pneumonia

_____ ketoacidosis and lactic acidosis are forms of this

_____ decreased respiratory activity, as in the case of drugs, can cause this

SELF TEST

Circle the correct answer to the questions below.

1. The two major fluid compartments of the body are the

   a. ICF and plasma.   b. ICF and interstitial fluid.   c. ICF and ECF.   d. ECF and plasma.   e. ECF and interstitial fluid.

2. The ECF and ICF

   a. contain equal amounts of sodium.   b. contain equal amounts of potassium. c. are equal in osmolarity.   d. are equal in volume.   e. more than one of the above is correct.

3. The hormone that promotes the loss of both fluid and electrolytes in the urine is

   a. ADH.   b. aldosterone.   c. ANF.   d. epinephrine.   e. insulin.

4. A person has taken a diuretic and as a result is producing large volumes of hypotonic urine. This would cause which of the following effects?

   a. increased plasma osmolarity   b. increased ICF osmolarity   c. decreased water concentrations in the plasma.   d. decreased water concentrations inside of the cells. e. more than one of the above is correct.

5. Proper water concentration in the body fluids is largely controlled by

   a. aldosterone.   b. ANF.   c. ADH.   d. the ANS.   e. none of the above.

6. Changes in the rate of sodium uptake or excretion do not affect the concentration of sodium in the ECF because

   a. potassium excretion or retention will compensate.   b. for every sodium retained or excreted, a hydrogen ion will be retained or excreted.   c. water osmotically follows the movement of sodium thereby stabilizing its concentration.   d. the fluid volumes remain constant.   e. none of the above.

7. A drop in ECF volume will result in

   a. increased ADH secretion.   b. increased aldosterone secretion.   c. decreased ANF secretion.   d. increased sodium retention.   e. more than one of the above is correct.

8. Increasing aldosterone concentrations

   a. increase potassium levels in the ECF.   b. increase potassium levels in the ICF. c. increase potassium levels in the plasma only.   d. decrease potassium levels in the ECF.   e. decrease potassium levels in the plasma only.

9. A substance that releases hydrogen ion in solution best defines a (an)

   a. acid.   b. base.   c. salt.   d. anion.   e. cation.

10. A person who had a plasma pH of 7.2 would be

    a. normal.   b. acidotic.   c. alkalotic.

11. Normal pH is maintained in the body by

    a. bicarbonate.   b. the kidney.   c. the respiratory system.   d. protein.   e. more than one of the above is correct.

12. The most important acid in the body is

    a. sulfuric acid.   b. phosphoric acid.   c. carbon dioxide.   d. lactic acid.   e. pyruvic acid.

13. Adding a strong acid such as hydrochloric acid to sodium bicarbonate converts the hydrochloric acid into

a. sulfuric acid.   b. carbonic acid.   c. lactic acid.   d. bicarbonate.   e. none of the above

14. The respiratory tract assists in regulating acid-base balance by

a. regulating the levels of metabolic acids.   b. regulating oxygen supply to the tissues.   c. regulating the carbon dioxide levels.   d. excreting bicarbonate.   e. excreting hydrogen ion.

15. Metabolic alkalosis is usually caused by

a. emphysema.   b. cardiac arrest.   c. congestive heart failure.   d. loss of bicarbonate.   e. severe vomiting.

CHAPTER 28
The Reproductive System

OVERVIEW

The male and female reproductive systems are the only one of the major systems which are not essential to the life of the individual. Ironically, they are the most essential organ systems to the species because they are the ones that insure the human species will be perpetuated. They are also the organ systems in which the greatest dichotomy exists between males and females. This is due to the fact that in female, care and nourishment of the developing fetus is a function which is not paralleled in males.

This chapter introduces you to the structure and function of the reproductive systems in both males and females. Because of the dichotomy in both structure and function, each sex is considered separately. Males are first considered, beginning with the anatomy of the male tract, and then considering male sexual function and the role of hormones. The female is then considered, first the anatomy and then the physiology. Physiological considerations in females include the ovarian cycle, menstrual cycle, and finally pregnancy. Physiology in females is much more complex than in males. This is largely due to the complex coordination between the ovarian and menstrual cycles which are necessary to insure fertility and conception.

CHAPTER OUTLINE

    A. Introduction
    B. The reproductive system of the male
       1. Male reproductive anatomy
          a. The testes
             (1) The descent of the testes

        (2) Historical organization
     b. The anatomy of a spermatozoan
     c. The male duct system
        (1) The epididymis
        (2) ductus deferens
        (3) The urethra
     d. The penis
     e. The accessory glands
        (1) The seminal vesicles
        (2) The prostate gland
        (3) The bulbourethral glands
     f. Semen
  2. Male reproductive physiology
     a. Male sexual function
     b. The role of hormones
C. The reproductive system of the female
  1. Female reproductive anatomy
     a. The ovaries
     b. The uterine tubes
     c. The uterus
     d. The vagina
     e. The external genitalia
     f. The mammary glands
  2. Female reproductive physiology
     a. Female sexual function
     b. The ovarian cycle
        (1) Hormones and the ovarian cycle
     c. The menstrual cycle
        (1) Menses
        (2) The proliferative phase
        (3) The secretory phase
        (4) Hormones and the menstrual cycle
     d. Pregnancy, hormones and maternal systems
     e. Aging and menopause
D. Hormones and reproductive success
E. Clinical patterns
  1. Developmental abnormalities
  2. Physiological problems
  3. Trauma, infection, and disease
  4. Technology and treatment of infertility

LEARNING ACTIVITIES

Complete each of the following items by supplying the appropriate word or phrase.

1. A fertilized egg is known as a _____.

2. The _____ refers to the external genitalia of either sex.

3. The path that a spermatozoa follows would be as follows.

    testes - (a)_____ - (b)_____ _ (c)_____ -urethra

4. The accessory glands of the male reproductive system include the _____, _____, and the _____ glands.

5. The testes are enclosed in a fleshy pouch, the _____.

6. The _____ is a layer smooth muscle found in the dermis of the scrotum.

7. A layer of skeletal muscle, the _____ muscle lies beneath the dermis of the scrotum.

8. Contraction of the _____ tenses the scrotum and pulls the testes closer to the body.

9. The testes descend into the scrotum through the _____ canals in the body wall.

10. Match the term listed below with the appropriate descriptive phrase that follows.

a. tunica albuginea b. septa c. lobules d. seminiferous tubules e. rete teste
f. efferent ducts g. interstitial cells h. spermatocytes i. spermatids j. sustentacular cells

_____ cells derived from the spermatogonia

_____ cells which form testosterone

_____ where spermatogenesis occurs

_____ a maze formed by the interconnections of the straight tubules

_____ the connections between the rete teste and the epididymis

_____ partitions formed by invaginations of the tunica albuginea

_____ subdivisions, formed by the septa, where the seminiferous tubules are located

_____ a dense connective tissue capsule that surrounds each teste

_____ control what gets into the seminiferous tubules and also secrete inhibin

_____ small cells formed by the division of the spermatocytes

11. The three parts of a sperm are the _____, _____, and _____.

12. The _____ cap of the sperm contain enzymes necessary for fertilization.

13. The mitochondria of a sperm are contained in the _____.

14. The spermatozoan tail is the only _____ found in the human body.

15. The _____ monitors and adjusts the tubular fluid, maintaining a suitable environment for sperm maturation, and it acts as a recycling center for damaged or deceased spermatozoa.

16. The epididymis is a tube which is around _____ meters long.

17. The epididymis can be divided into a _____, _____, and _____.

18. The epididymis connects to the _____.

19. Label figure 28.1, the testis and epididymis, using the terms listed below. Color each part. Refer to figure 28.3 in your textbook for assistance.

vas deferens   epididymis   nerves and blood vessels
spermatic cord   tunica albuginea   seminiferous tubules
septum   rete testis   efferent ducts

20. The enlarged and expanded portion of the vas deferons just before it reaches the prostate gland is the _____.

21. The confluence of the empulla and the seminal vesicle forms the _____ duct.

22. The _____ urethra passes through the center of the prostate gland, the _____ urethra penetrates the urogenital diaphragm, and the _____ urethra is contained within the penis.

Labels: vas deferens   epididymis   nerves and blood vessels   spermatic cord   tunica albuginea
seminiferous tubules   septum   rete testis   efferent ducts

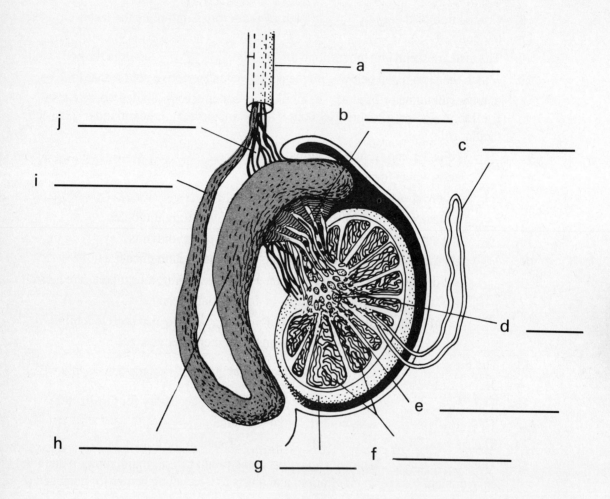

a _____

b _____

c _____

j _____

i _____

d _____

e _____

h _____

g _____   f _____

Figure 28.1

23.   The external urethral meatus of the penis is surrounded by the __a__, which may be
covered by a flap of skin known as the __b__. This flap of skin has its origin at the
narrow __c__ of the penis. Internally the body is made up largely of three cylinders
of erectil tissue. There are two __d__ which are located dorsal and lateral to the
urethra, and a single __e__ which surrounds the urethra.

a. _____

b. _____

c. _____

d. _____

e. _____

24. Match the gland listed below with the appropriate descriptive statement which follows.

a. seminal vesicles   b. prostate gland   c. bulbouretharal glands

_____ produce a solution which is rich in fructose

_____ fluid produced by this gland comprises about 30 percent of the volume of the semen

_____ glands situated at the base of the penis

_____ gland that surrounds the the proximal portion of the urethra after it exits the bladder

_____ produce 60 percent of the semen

_____ fluid from this gland is very alkaline and neutralizes acidity both in the urethra and the female tract

_____ produce a thick, alkaline fluid that neutralizes and lubricates the urethra

25. The normal sperm count is usually between _____ and _____ million per ml of semen.

26. A normal ejaculate will consist of about _____ ml of semen.

27. _____ is an enzyme which has been isolated from semen that seems to have antibacterial properties.

28. The four phases of male sexual function are _____, _____, _____, and _____.

29. Erection is brought about by _____ nervous activity.

30. _____ is caused by the rhythmetic contractions of smooth muscle throughout the male tract.

31. Ejaculation is caused by the contraction of the _____ and _____ muscles.

32. Label figure 28.2, the male reproductive tract, using the terms listed below. Color each part of the tract. Refer to figure 28.1 in your textbook for assistance.

ureter	bladder	seminal vesicle	ejaculatory duct
prostatic urethra	epididymis	penile urethra	glans
corpus cavernosum	prepuce	urethral meatus	scrotum
bulbourethral gland	testis	urogenital diaphragm	
anus	prostate gland		

33. Label figure 28.3, cross section of the penis, using the terms listed below. Color each structure. Refer to figure 28.8 in your textbook for assistance.

superficial dorsal vein	deep dorsal vein	median septum
penile urethra	corpus spongiosum	fibrous coat
corpus cavernosum	dorsal artery	

34. _____ is the draining of the blood from the penis following ejaculation, and is mediated by sympathetic activity.

35. _____ causes secretion of testosterone by the interstitial cells.

36. Functional maturation of spermatozoa and secondary sex characteristics of males are determined by _____.

37. Spermatogenesis is stimulated by _____ and inhibited by _____.

Label: ureter bladder seminal vesicle ejaculatory duct prostatic urethra epididymis
        penile urethra glans corpus cavernosum prepuce urethral meatus scrotum
        bulbourethral gland testis urogenital diaphragm anus prostate gland

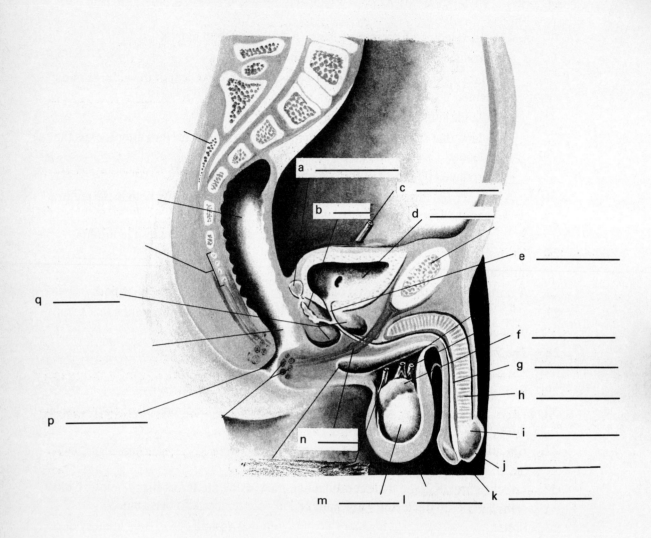

Figure 28.2

38. The principal organs of the female reproductive system are the _____,
    _____, _____, and _____.

39. The ovaries, uterine tubes, and uterus are enclosed within an mesentery known as
    the _____.

40. The _____ is a thickened fold of mesentery that stabilizes and sup-
    ports each ovary.

41. Each ovary is surrounded by a fibrous capsule, the _____, and con-
    sists of an outer _____ and an inner _____.

42. The expanded end, the _____, of the uterine tubes is divided into
    fingerlike projections known as the _____.

Label: superficial dorsal vein   deep dorsal vein   median septum   penile urethra   corpus spongiosum
         fibrous coat   corpus cavernosum   dorsal artery

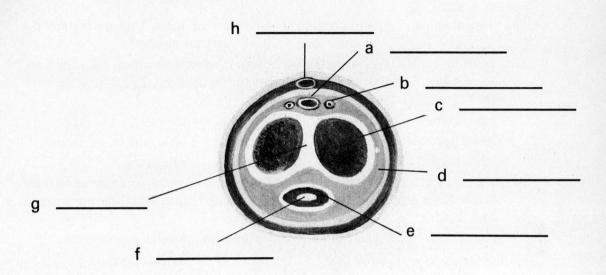

Figure 28.3

43. The epithelium that lines the uterine tubes is covered with _____.

44. It normally takes _____ days for an ovum to transit the uterine tubes to the uterus.

45. The _____ provides mechanical protection and nutritional support for the developing embryo.

46. Besides the broad ligament, there are three sets of ligaments which stabilize and support the uterus. They are the _____, _____, and _____ ligaments.

47. The three regions of the uterus are the _____, _____, and _____.

48. The three layers of the uterus consist of an inner lining, the _____, a muscular middle layer, the _____, and an outer _____.

49. It is the _____ layer of the endometrium that undergoes changes during the menstrual cycle.

50. The _____ is a muscular tube extending from the cervix of the uterus to the external genitalia.

51. The vagina serves to receive the _____ during coitus and also as the birth canal.

52. Match the terms below with the appropriate descriptive statements that follow.
    a. vulva   b. vestibule   c. labia minora   d. clitoris   e. prepuce   f. mons pubis
    g. labia majora
    _____ mound of fatty tissue that forms the anterior border of the vulva
    _____ the collective name for the female external genitalia
    _____ female equivalent of the penis
    _____ encircle and partially conceal the labia minora
    _____ an extension of the labia minora that encircles the clitoris

_____ common area that both the vagina and urethra open into

_____ folds that bound the vestibule

53. The reddish brown region that surrounds each nipple of the breast is the _____.

54. The mammary glands consist of a number of lobes that drain into the _____ ducts that eventually reach the nipples.

55. Label figure 28.4, sagittal section of the female reproductive tract, using the terms listed below. Color each structure. Refer to figure 28.10 in your textbook for assistance.

uterine tube   fimbriae   uterus   cervix   anus
round ligament   pubic symphysis   clitoris   labia minora   ovary
labia majora   bladder   urethra   vagina

56. Label figure 28.5, the uterus, ovary, and associated structures using the terms listed below. Color each structure. Refer to figures 28.10, 28.11, and 28.12 in your textbook for assistance.

cervix   vagina   round ligament   broad ligament body
of the uterus   uterine tube   ampulla   ovarian ligament
ovary fimbriae

57. Label figure 28.6, the female external genitalia, using the terms listed below. Color each structure. Refer to figure 28.13 in your textbook for assistance.

prepuce   mons pubis   clitoris   urethra   anus
labia minora   labia majora   vagina   vulva

58. During female sexual arousal, _____ activation leads to engorement of the erectile tissues in the clitoris.

59. The maturation of ova or eggs occurs on a _____ basis.

60. Ovum development occurs in ovarian _____.

61. An oocyte which is enclosed by a single layer of epithelial cells constitutes a _____ follicle.

62. In _____ follicles, the follicular cells enlarge and several layers form around the oocyte.

63. The space that forms between the oocyte and the follicle cells in a growing follicle is the _____.

64. The _____ that form on the surface of the oocyte and the follicular cells in the zona pellucida increase the surface area contact by about 35 percent.

65. When the follicle walls begin to thicken and some of the cells begin secreting fluid, a _____ follicle has been formed.

66. In a _____ follicle, there is a large fluid filled chamber termed the antrum.

67. Following ovulation, the layer of follicle cells surrounding the ovum is termed the _____.

68. The follicular cells of the ruptured follicle proliferate and form a structure known as the _____.

69. As the corpus luteum degenerates, fibroblasts invade and form a scar tissue known as the _____.

70. The events from primoridal follicle to corpus albicans constitute the _____ cycle.

71. The stimulus for follicle maturation to begin is the hormone _____.

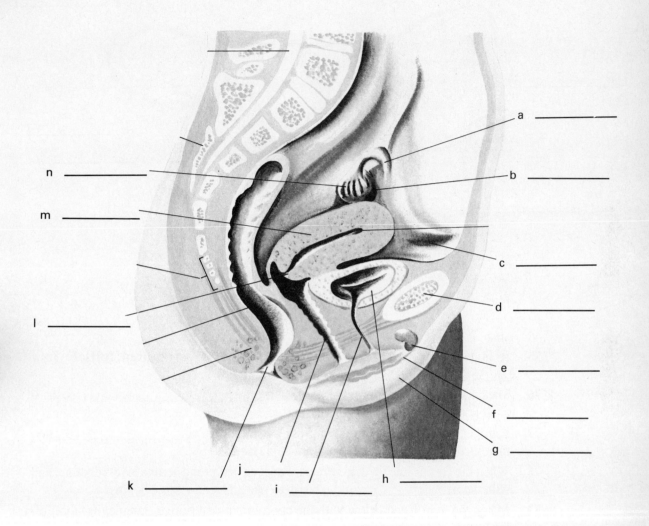

Figure 28.4

72. As the follicle begins to mature, the follicular cells produce _____.

73. Estrogen inhibits the releases of _____ but stimulates the release of _____.

74. High levels of _____ cause the rupturing of the mature follicle and ovulation.

Label: cervix vagina round ligament broad ligament body of the uterus uterine tube ampulla ovarian ligament ovary fimbriae

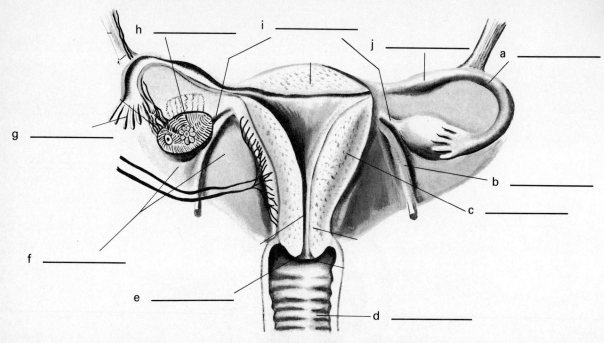

Figure 28.5

75. It is LH that stimulates the conversion of the ruptured follicle into _____.

76. The corpus luteum produces estrogens and _____, which is the principal hormone of the postovulatory period.

77. Estrogens and progesterones suppress the secretion of _____ and therefore both FSH and LH levels are depressed.

78. When the corpus luteum degenerates, estrogen and progesterone levels decline, and with them, _____ begin to rise and a new cycle begins.

79. Match the term listed below with the appropriate descriptive statement(s) that follow.

a. menses b. PMS c. proliferative phase d. secretory phase e. menarche
f. menopause g. blastocyst h. implantation i. HCG j. placenta k. relaxin
l. HPL m. prolactin n. colostrum

_____ combined with HPL it stimulates the production of milk by the mammary glands during pregnancy

_____ hormone produced by the placenta which causes the pubic symphysis to soften and become more flexible

_____ first hormone that is produced by the developing placenta

_____ organ that provides nutrition and support for the embryo

_____ period of time when the menstrual cycle begins

_____ period of time when the menstrual cycle terminates

_____ phase of the cycle that is largely controlled by progesterone

_____ phase also known as the postovulatory, or luteal phase

_____ major stimulation for this phase, when the endometrium is proliferating, is estrogen

Label: prepuce mons pubis clitoris urethra anus labia minora labia majora vagina vulva

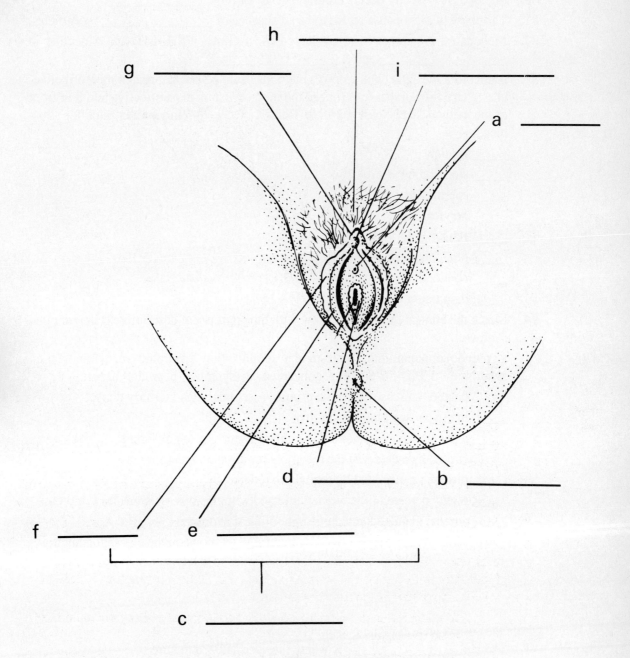

h _____

g _____   i _____

a _____

d _____   b _____

f _____   e _____

c _____

Figure 28.6

_____ phase of the menstrual cycle which is also known as the preovulatory, or follicular phase

_____ period when the endometrium sloughs off and bleeding occurs

_____ represents the begining of a new menstrual cycle

_____ placental hormone that aids in stimulating milk production

_____ a series of physiological and psychological changes that occur in some women prior to the menstruation period

_____ hollow ball of cells that constitutes the embryo upon its arrival at the uterus

_____ mechanism by which the embryo embeds itself in the endometrium

_____ the secretion of the mammary glands during the first two or three days of the infant's life

80. Menopause typically occurs around the age of _____.

81. The cause of menopause seems to be an absence of _____ follicles.

82. Menopause is characterized by increases in GnRF, FSH, LH, and a decline in _____ and _____.

83. Rank order the eight contraceptive methods listed below in terms of their effectiveness. Place the number 1 in front of the method that is most effective and 8 in front of the method that is least effective. Refer to Table 28.3 in your textbook for assistance.

_____ rhythm

_____ sterilization

_____ spermicide

_____ cervical cap

_____ IUD

_____ oral contraceptive

_____ condom

_____ diaphragm

84. Match the clinical term listed below with the appropriate descriptive statement that follows.

a. pseudohermaphrodite   b. testicular feminization syndrome   c. adrenogenital syndrome   d. PID   e. STD   f. gonorrhea   g. syphilis   h. genital herpes

_____ bacterial disease that has a primary, secondary, and tertiary phase

_____ major cause of female sterility

_____ a person whose anatomical sex and genetic sex are different

_____ caused by a defect in the receptors for circulating androgens

_____ diseases transmitted by sexual intercourse

_____ eighty percent of the women infected with this disease show no symptoms

_____ adrenal glands secrete large quantities of androgens

_____ viral disease characterized by painful ulcerations on the external genitalia

SELF TEST

Circle the correct answer to the questions below.

1. The _____ is where the spermatozoa mature.

   a. vas deferens   b. seminiferous tubules   c. ejaculatory ducts   d. epididymis   e. seminal vesicle

2. Spermatogenesis is initiated by

   a. LH.   b. FSH.   c. testosterone.   d. estrogen.   e. progesterone.

3. The _____ produce a fluid that is rich in fructose.

   a. testes   b. seminal vesicles   c. prostate gland   d. bulbourethral glands   e. epididymis

4. The _____ contributes about 30 percent of the volume of the semen.

   a. testes   b. epididymis   c. seminal vesicle   d. prostate gland   e. bulbourethral gland

5. The female gametes are produced in the _____ of the ovary.
    a. tunica albuginea   b. medulla   c. lobules   d. cortex   e. epididymis

6. The funnel shaped region of each uterine tube is the
    a. broad ligament.   b. infundibulum.   c. isthmus.   d. ampulla.   e. fimbriae.

7. The lining of the uterus is the
    a. myometrium.   b. serosa.   c. endometrium.   d. cervix.   e. fundus.

8. The female equivalent of the penis is the
    a. clitoris.   b. labia majora.   c. labia minora.   d. mons pubis.   e. urethra.

9. The mature follicle is known as the _____ follicle.
    a. primordial   b. primary   c. secondary   d. tertiary   e. quaternary

10. The follicular cells that surround the ovum following ovulation form the
    a. antrum.   b. zona pellucida.   c. corona radiata.   d. secondary follicle.   e. none
    of the above.

11. When the follicle begins to develop, the follicle cells produce
    a. FSH.   b. HCG.   c. LH.   d. estrogen.   e. progesterone.

12. Proliferation of the endometrium during the proliferative phase is stimulated by
    a. progesterone.   b. estrogen.   c. LH.   d. testosterone.   e. the corpus luteum.

13. The first hormone produced by the placenta is
    a. prolactin.   b. HCG.   c. FSH.   d. estrogen.   e. progesterone.

14. The corpus luteum produces
    a. LH.   b. FSH.   c. estrogen.   d. progesterone.   e. estrogen and progesterone.

15. The STD characterized by ulcerative lesions on the genitals is
    a. gonorrhea.   b. syphilis.   c. genital herpes.   d. PID.   e. psoriasis.

CHAPTER 29
Development and Inheritance

OVERVIEW

In the previous chapter you learned about the structure and function of the reproductive system. This chapter is really a continuance of this discussion, except now the focus is on the end product of these systems, the new individual.

The chapter begins with a discussion of the formation of the sex cells by the respective reproductive systems, and their union during the process of fertilization. Following conception, the discussion moves into a consideration of the relationship between genetics and the developmental process. The next major topic to be considered is the prenatal development of the offspring, in which the principal events of the three trimesters of development are examined. The events of birth are then described and the remainder of the chapter is taken up with a consideration of postnatal development through the final stages of life.

It needs to be emphasized that the developmental process which begins at conception does not stop with birth. Rather the entire life of each individual, until ultimately death, is a series of physiological and anatomical changes that are part of this continuous process.

CHAPTER OUTLINE

A. Introduction
B. Meiosis and the formation of gametes
   1. Fertilization
C. Genetics, Development, and Inheritance
   1. Genes and chromosomes
      a. Autosomal chromosomes

        b. Sex chromosomes
    2. Induction and the regulation of development
        a. Competence and developmental timing
  D. Prenatal development
    1. The first trimester
        a. Cleavage and blastocysts formation
        b. Implantation
        c. Placentation
            (1) The extraembryonic membranes
            (2) Placental formation and growth
        d. Embryogenesis
    2. The second and third trimesters
        a. Pregnancy and maternal systems
  E. Labor and delivery
    1. Common problems with labor and delivery
        a. Breech births
        b. Premature delivery
        c. Multiple births
  F. Postnatal development
    1. Infancy and childhood
        a. Monitoring postnatal development
    2. Adolescence and maturity
  G. Death and dying
  H. Clinical patterns
    1. Inherited disorders
        a. Simple inheritance
        b. Multifactorial inheritance
        c. Chromosomal abnormalities
        d. Autosomal chromosomes
        e. Chromosomal analysis
        f. Sex chromosome abnormalities
        g. Mutations
    2. Teratogens and abnormal development
        a. Common teratogenic stimuli: alcohol and smoking

LEARNING ACTIVITIES

Complete each of the following items by supplying the appropriate word or phrase.

1. The process of _____ regulates the appearance and modification of physical and physiological characteristics throughout life.

2. _____ is the creation of different cell types by the turning on and off of selective genes.

3. Development begins at _____, or conception.

4. The first two months of development following fertilization is _____.

5. From the ninth week to birth constitutes _____ development.

6. _____ development begins at birth and continues throughout life.

7. Development involves the division and _____ of cells, and changes in genetic activity that produce and modify anatomy and physiology.

8. _____ refers to the transfer of genetically determined characteristics from generation to generation and _____ is the study of the mechanisms of this transferance.

9. Match the term listed below with the appropriate descriptive statement(s) which follow.

a. meiosis b. diploid c. haploid d. synapsis e. tetrad f. chromatids g. reductional division   h. equational division

_____ the number of chromosomes that a somatic cell will contain

_____ first division of meiosis produces these kind of cells

_____ the second division of meiosis

_____ pairing of homologous (maternal and paternal) chromosomes

_____ special kind of cell division in which the diploid chromosome number is reduced to the haploid number

_____ name that is given to the duplicated chromosome strands that are held together by centromeres

_____ first division of meiosis

_____ name given to the maternal and paternal pairs if chromosomes when they are synapsed because of the appearance of the four chromatid strands

10. A diploid cell contains 60 chromosomes. Following meiosis of this cell, each daughter cell will have _____ chromosomes.

11. Spermatogonia give rise to _____.

12. Primary spermatocytes undergo _____ division to give rise to secondary spermatocytes.

13. Secondary spermatocytes undergo _____ division to give rise to spermatids.

14. In terms of chromosome numbers, primary spermatocytes are _____ cells and secondary spermatocytes are _____ cells.

15. Oogonia give rise to _____.

16. A primary oocyte undergoes reductional division and gives rise to a secondary oocyte and a _____.

17. A secondary oocyte undergoes equational division and gives rise to an ovum and a _____.

18. For each spermatogonium that undergoes meiosis, _____ haploid spermatozoa will be formed.

19. For each oogonium that undergoes meiosis, one mature haploid ovum will be formed and _____ polar bodies.

20. Label figure 29.1, gametogenesis, using the terms listed below. Refer to figure 29.2 in your textbook for assistance.

primary spermatocyte   secondary spermatocyte   spermatids
spermatozoa   primary oocyte   secondary oocyte
first polar body   second and third polar bodies   ovum

21. The reason that the ovum is so much larger than a sperm is that it must provide _____ for the developing embryo for a week following fertilization.

22. Fertilization occurs in the _____ of the female tract.

23. The intercellular cement that holds the cells of the corona radiata together are broken down by the enzyme _____ which is released by the acrosomal cap of the sperm.

GAMETOGENESIS

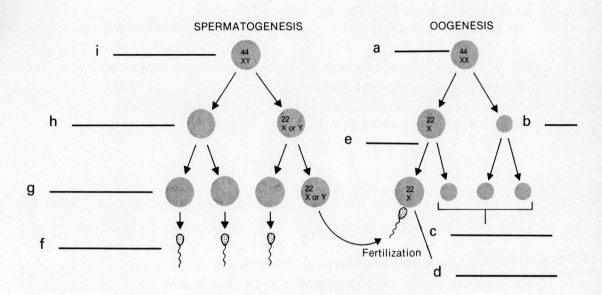

Figure 29.1

24. Once a sperm has penetrated the secondary oocyte, it becomes _____.

25. Inside of the ovum the female pronucleus and the male pronucleus fuse in a process called _____.

26. The component genes of an individual determine the _____, and the visible traits constitute the _____.

27. A maternal and paternal chromosome which contain genes that govern the same traits are termed _____.

28. Genes that control the same traits but are located on separate chromosomes are known as _____.

29. If a pair of alleles on homologous chromosomes are identical, that is, if they contain the same instructions, then the individual is said to be _____ for the trait governed by those alleles. If the alleles carry different instructions the individual is said to be _____.

30. A _____ allele will be expressed in the phenotype regardless of the nature of the alternate allele.

31. In order for a _____ allele to be expressed it must be present in a double dose, that is, homozygous.

32. If a man contains the alleles A and a (Aa) for a given trait, each sperm produced by him will contain the allele _____ or _____, but not both because of meiosis.

33. If you know the alleles of the parents, it is possible to predict the various genetic combinations for their children using a device known as a _____ square.

34. Humans have 22 pairs of homologous _____ and one pair of _____ chromosomes.

35. A normal male will have the genotype _____ and a normal female will be _____.

36. Females can produce ova that will only have the _____ sex chromosome while males can produce sperm that will have the _____ or _____ sex chromosome.

37. While the Y chromosome seems to be genetically inert, the X chromosome has a full complement of genes that determine _____ traits.

38. A woman with normal vision who had a color blind father marries a normal man. List below the possible genotypes of the sons and daughters produced and indicate which, if any, would be color blind.

Sons

_____   _____

Daughters

_____   _____

39. _____ is the alteration of a cell's activities during development by chemical substances released by nearby cells.

40. Induction usually involves the turning on or off of certain _____ in the cell begin acted upon.

41. The four major groups of events that occur during the first trimester of pregnancy are _____, _____, _____, and _____.

42. _____ is a series of mitotic divisions that subdivides the cytoplasm of the zygote.

43. Each cell formed by cleavage is termed a _____, and after a period of time the zygote is converted into a ball of cells termed a _____.

44. The stage following the morula is known as the blastocyst and is a hollow ball of cells with a central cavity termed the _____.

45. The outer layer cells of the blastocyst is the _____, while an inner group of cells at one end is known as the _____ cell mass.

46. The process by which the blastocyst enters the endometrium of the uterus is known as _____.

47. At about the time of implantation of the trophoblast, the inner cell mass has undergone a process termed _____ which results in the formation of the three embryonic tissue layers.

48. Besides the various organ systems of the body, the three germ layers give rise to the _____ membranes.

49. The four extraembryonic membranes are the _____, _____, _____, and the _____

50. The yolk sac is derived from _____ and _____.

51. The yolk sac does not provide yolk for the embryo as nutrition is taken care of by the placenta, but does serve as an important source of _____ cells.

52. The amnion is derived from the _____ and _____.

53. The amnion encloses _____ that cushions the fetus.

54. The base of the allantois eventually gives rise to the _____.

55. The last and outermost of the extraembryonic membranes, the _____ is formed from a layer of mesoderm that combines with the trophoblast.

56. The mesoderm extends along the core of each trophoblastic villus to form _____ villi which will be the major connecting link of the placenta to the mother.

57. As the chorion expands into the lumen of the uterus, the endometrial connection becomes progressively less. At the end of placental formation, only disc shaped area of endometrial contact, called the _____, still provides an exchange surface.

58. The chorionic villi provide an exchange surface of about _____ square meters.

59. By the start of the second trimester, rudiments of all of the major _____ have formed.

60. It is during the _____ that all of the organ systems become functional.

61. Label figure 29.2, the extraembryonic membranes, using the terms listed below. Color each structure. Refer to figure 29.9 in your textbook for assistance.

chorionic villus	allantois	yolk sac	umbilical cord
embryo	amniotic cavity	amnion	chorion

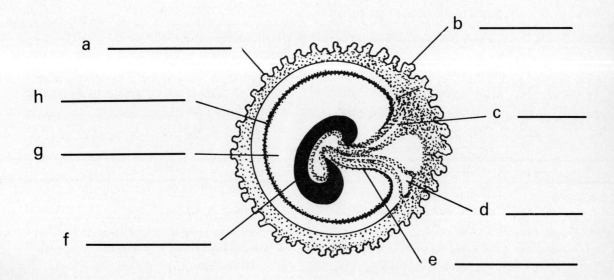

Figure 29.2

62. For each of the events of development which are listed below, place a number representing the month of prenatal development in which it begins. Refer to Table 29.3 in your textbook for assistance.

_____ central nervous system tract formation

_____ degeneration of the embryonic kidneys

_____ formation of the eye and the ear

_____ nail and hair formation

_____ descent of the testes

_____ heartbeat

_____ articulations

_____ epiphyseal plates

_____ myelination of the spinal cord

63. By the end of gestation, maternal blood volume has increased by _____ percent and the GFR has risen by _____ percent.

64. _____ is the expulsion of the fetus.

65. The contractions of the uterus that result eventually in fetal expulsion are termed _____.

66. Labor is divided into three stages, _____, _____, and _____.

67. In the dilation phase, the _____ dilates completely.

68. Late in the dilation phase, the _____ usually ruptures.

69. During the _____ stage, the fetus is being pushed through the vagina.

70. The final stage of labor, the _____ stage, occurs after delivery and consists of powerful uterine contractions that expel the placenta as the "afterbirth."

71. A _____ birth is when the legs or buttocks enter the vagina as opposed to the head of the infant.

72. _____ twins result from separate fertilizations of two ova.

73. _____ twins result from separation of the inner cell mass into two parts prior to gastrulation.

74. If the separation of blastomeres or the inner cell mass is not complete, twins may be born joined to one another, a condition referred to as _____ (Siamese) twins.

75. The major stages of postnatal development include _____, _____, _____, _____, and _____.

76. The _____ stage extends from birth to one month of age.

77. _____ lasts until age two and _____ runs until puberty.

78. The dividing line between adolescence and maturity is hazy but is usually considered to be the age at which _____ has ceased.

79. List the four major processes that seem to play a major role in aging.

a. _____

b. _____

c. _____

d. _____

80. The leading cause of death for both males and females under age 34 is _____.

81. Match the clinical condition listed below with the appropriate descriptive statement which follows.

a. Tay-Sachs disease   b. coronary artery disease   c. Down's syndrome
d. Klinefelter's syndrome   e. Turner's syndrome   f. teratogen   g. FAS

_____ abnormality caused by a third twenty-first chromosome

_____ an example of a disease which has a multifactorial inheritance pattern

_____ substance that will cause developmental abnormalities

_____ disease that results from simple inheritance

_____ individuals with a chromosome pattern of XXY

_____ caused by consumption of alcohol by a pregnant mother

_____ females having a genotype of X0, meaning they lack a second X chromosome

SELF TEST

Circle the correct answer to the questions below.

1. The transformation of embryonic cells into specialized types best describes

a. meiosis. b. spermatogenesis. c. oogenesis. d. differentiation. e. embryogenesis.

2. Which of the following cells is diploid?

a. secondary spermatocyte   b. secondary oocyte   c. first polar body   d. primary spermatocyte   e. spermatid

3. Two chromosomes that contained genes that regulated the same traits would be described as

a. identical. b. homologous. c. analogous. d. alleles. e. none of the above

4. An individual who had two identical alleles for a trait would be _____ for that trait.

a. homologous   b. heterozygous   c. homozygous   d. autosomal   e. recessive

5. In order for a recessive gene to be expressed in the phenotype, it must be

a. heterozygous. b. homozygous. c. sex linked. d. autosomal linked. e. more than one of the above is correct.

6. In order for a girl to be born color blind

a. her father must have been normal and her mother color blind.   b. her father must have been color blind and her mother completely normal.   c. her father must have been color blind and her mother must also have been color blind.   d. her father must have been color blind and her mother had to carry the color blind gene on at least one X chromosome.   e. none of the above.

7. The turning on or off of specific genes by environmental factors is known as

a. organogenesis. b. differentiation. c. induction. d. competence. e. gestation.

8. The cells that compose the morula are known as

a. blastomeres. b. trophoblasts. c. lacunae. d. blastocysts. e. hypoblast.

9. That part of the blastocyst that actually becomes the embryo is the

a. trophoblast. b. blastocoel. c. inner cell mass. d. chorion. e. amnion.

10. The extraembryonic membrane that gives rise to blood cells is the

    a. yolk sac. b. allantois. c. chorion. d. amnion. e. endoderm.

11. The extraembryonic membrane which functions as the primary placental transfer membrane between mother and child is the

    a. amnion. b. allantois. c. chorion. d. amnion. e. ectoderm.

12. All of the organ rudiments have been established by the end of

    a. gestation. b. the first trimester. c. the second trimester. d. the third trimester. e. implantation.

13. A major source of stimulus for the contraction of the uterus is

    a. progesterone. b. LH. c. FSH. d. relaxin. e. oxytocin.

14. The neonatal stage extends from birth to

    a. puberty. b. one month. c. 6 months. d. 7 years. e. maturity.

15. A person suffering from Klinefelter's syndrome would have the which of the following chromosome genotypes?

    a. XYY b. X0 c. YO d. XXY e. XXX

ANSWERS

Chapter 1. 1. respond, adaptability, grow, reproduce, movement, locomotion;
2. life; 3. anatomy; 4. function; 5. i, g, h, a, d, b, c, e; 6. cytology;
7. histology, organology; 8. structure, function; 9. 4, 3, 2, 1, 5, 6, 7;
10. homeostasis; 11. receptor; 12. effector; 13. a. stimulus, b. receptor,
c. effector; 14. negative; 15. positive; 16. disease; 17. disease;
18. development; 19. embryology; 20. congenital defects; 21. c, a, b, d, e;
22. hypothesis; 23. testable, unbiased, repeatable; 24. theory;
25. predictions; Self Test 1. e; 2. e; 3. c; 4. b; 5. d; 6. a;
7. c; 8. d; 9. a; 10. b; 11. d; 12. b; 13. c; 14. c; 15. b.

Chapter 2. 1. mass; 2. atom; 3. protons, neutrons, electrons; 4. c, f, a,
b, h, g, d, e; 5. 2, 8; 6. 23; 7. neutrons; 8. radioisotopes; 9. outer;
10. sharing, gaining, losing; 11. outer; 12. compound; 13. covalent;
14. double; 15. polar; 16. ionic; 17. ions; 18. cations, anions;
19. electrical; 20. valence; 21. positive; 22. nitrogen, oxygen;
23. shape; 24. 18; 25. synthesis, decomposition, exchange, decomposition,
synthesis; 26. exothermic, endothermic; 27. carbon, hydrogen; 28. polar;
29. hydrogen; 30. solute; 31. dissociate; 32. electrolytes; 33. pH;
34. hydrogen ion; 35. 7; 36. decreases; 37. ionizes; 38. base;
39. salts; 40. buffers; 41. carbon, hydrogen; 42. metabolism;
43. catabolism, energy; 44. anabolism; 45. covalent bonds; 46. four;
47. hydrophobic, hydrophilic; 48. 1:2:1; 49. 3, 4, 5, 6, 7; 50. glucose;
51. disaccharide; 52. dehydration; 53. water, hydrolysis;
54. polysaccharides; 55. glycogen; 56. lipids; 57. carboxyl;
58. unsaturated; 59. essential fatty; 60. diglyceride; 61. fat; 62. energy;
63. prostaglandins; 64. cholesterol; 65. membranes; 66. cholesterol,

phospholipid, glycolipid;   **67.** amino acids;   **68.** 20;   **69.** peptide;
70. tripeptide;   **71.** fibrous;   **72.** globular;   **73.** b, a, d, c, c and d;
74. support, movement, transport, buffers, enzymes, coordination and control,
defense;   **75.** energy of activation;   **76.** catalyst;   **77.** enzymes;
78. substrates;   **79.** active site;   **80.** cofactors;   **81.** coenzymes;
82. enzyme;   **83.** glycoproteins, proteoglycans;   **84.** DNA, RNA;   **85.** sugar,
phosphate;   **86.** ribose, deoxyribose;   **87.** adenine, thymine, cytosine, guanine;
88. adenine, cytosine, guanine, uracil;   **89.** phosphate, sugar;   **90.** double;
91. T- A- G - C;   **92.** ATP;   **93.** ADP;   **94.** vitamins;   **95.** heme.   Self
Test **l.** 4;   **2.** b;   **3.** d;   **4.** c;   **5.** d.;   **6.** e;   **7.** b;   **8.** a;   **9.** a;
10. b;   **ll.** d;   **12.** c;   **13.** b;   **14.** b;   **15.** c.

Chapter 3. **l.** cytology;   **2.** 1000X;   **3.** transmission, scanning;   **4.** a. cilia
b. centriole   c. Golgi complex   d. smooth ER   e. mitochondrion   f. ribosomes
g. rough ER   h. nucleus   i. nucleolus   j. cell membrane   k. cytosol
l. microvilli;   **5.** cilia, cell membrane, nucleus, cytosol, lysosomes, mitochondria,
nucleolus, centrosome, ribosomes, Golgi apparatus, endoplasmic reticulum,
microvilli;   **6.** center;   **7.** integral;   **8.** pores;   **9.** microfilaments,
microtubules;   **10.** basal body;   **ll.** tubulin;   **12.** cilia;   **13.** cristae;
14. diffusion;   **15.** size, charge, lipid solubility, carrier molecules;   **16.** carrier;
17. sodium;   **18.** osmosis;   **19.** hypertonic, hypotonic, isotonic, hypotonic,
hypertonic, isotonic;   **20.** filtration;   **21.** active transport, energy;   **22.** carrier,
ATP;   **23.** pinocytosis, phagocytosis;   **24.** positive, negative;   **25.** potential
difference;   **26.** resting;   **27.** contraction;   **28.** metabolism;   **29.** secretion;
30. catabolism;   **31.** phosphorylation;   **32.** GTP, CTP, UTP, creatine phosphate;
33. substrate;   **34.** substrate;   **35.** NAD, FAD, FMN, Coenzyme Q;
36. electron transport system;   **37.** cytochrome;   **38.** energy;   **39.** ATP;
40. oxygen;   **41.** water;   **42.** ATP;   **43.** See figure 3.20;
44. mitochondria;   **45.** Kreb's cycle;   **46.** See figure 3.21;   **47.** acetyl CoA,
carbon dioxide, hydrogen, NADH2, FADH2, GTP;   **48.** glycolysis;
49. aerobic;   **50.** pyruvic;   **51.** 2;   **52.** acetyl CoA;   **53.** 4, 2, 4, 4, 6, 18;
54. anaerobic;   **55.** lactic acid;   **56.** 18;   **57.** pyruvic acid;   **58.** beta
oxidation;   **59.** transamination;   **60.** deamination;   **61.** Kreb's cycle;
62. urea;   **63.** gluconeogenesis;   **64.** glycolysis;   **65.** linoleic acid,
arachindonic, linolenic;   **66.** 10;   **67.** DNA;   **68.** amino acid;   **69.** gene;
70. transcription;   **71.** codon;   **72.** translation;   **73.** anticodon;   **74.** amino acid;
75. GGG UUU AAA CCC, CCC AAA UUU GGG;   **76.** RNA polymerase;
77. histones;   **78.** nucleosomes;   **79.** negative feedback, inducers, repressors;
80. prophase, metaphase, anaphase, interphase, prophase, telophase, telophase,
telophase, prophase;   **81.** activation.   Self test **l.** c;   **2.** d;   **3.** a;
4. hypotonic;   **5.** b;   **6.** b;   **7.** b;   **8.** c;   **9.** e;   **10.** c;   **ll.** b;   **12.** c;
13. d;   **14.** e;   **15.** b.

Chapter 4. **l.** histology;   **2.** epithelia, connective, muscle, nerve;   **3.** free;
4. extracellular;   **5.** endocrine;   **6.** permeability, sensitivity, protection,
secretion;   **7.** tight junction;   **8.** proteoglycan;   **9.** intermediate;
10. hyaluronic acid;   **ll.** basement membranes;   **12.** microvilli;   **13.** layers,
cells;   **14.** simple;   **15.** stratified;   **16.** squamous;   **17.** cuboidal;
18. columnar;   **19.** mesothelium, endothelium;   **20.** a, a, b, c, d, b, e, f;
21. simple, compound;   **22.** mixed;   **23.** merocrine;   **24.** apocrine;

25. holocrine;   **26.** goblet cells;   **27.** A. cuboidal B. simple columnar
C. pseudostratified ciliated columnar D. simple squamous E. stratified squamous
F. transitional;   **28.** compound;   **29.** serous;   **30.** cells, ground substance,
fibers;   **31.** fixed, wandering;   **32.** fibroblasts;   **33.** fat;   **34.** melanocytes;
35. macrophages;   **36.** histamine, heparin;   **37.** collagenous, elastic, reticular;
38. areolar;   **39.** adipose;   **40.** tendons, ligaments, aponeuroses;
41. irregular;   **42.** elastic;   **43.** reticular;   **44.** blood, lymph;
45. chondroitin;   **46.** lacunae;   **47.** chondrocytes;   **48.** perichondrium;
49. hyaline, elastic, fibrous;   **50.** a. reticular fibers   b. fixed macrophage
c. plasma cell   d. red blood cells   e. blood vessel   f. fat cells   g. blood vessel
h. mast cell,   i. elastic fibers   j. free macrophage   k. collagen fibers   l. fibroblast
m. lymphocyte;   **51.** fibrocartilage;   **52.** hyaline;   **53.** elastic;
54. interstitial, appositional;   **55.** minerals;   **56.** osteocytes;   **57.** canaliculi;
58. mucous, serous, cutaneous;   **59.** mucous;   **60.** serous;   **61.** peritoneum;
62. parietal, visceral;   **63.** reduce;   **64.** cutaneous;   **65.** synovial;
66. synovial;   **67.** smooth, skeletal, cardiac;   **68.** striations;   **69.** cardiac;
70. smooth;   **71.** skeletal;   **72.** nervous;   **73.** neuron;   **74.** neuroglia;
75. axon, dendrite;   **76.** inflammatory;   **77.** histamine, heparin;   **78.** red;
79. swelling;   **80.** pain;   **81.** fibrin;   **82.** microphages;   **83.** neutrophil;
84. pus;   **85.** abscess;   **86.** cyst.   Self Test. **1.** a;   **2.** c;   **3.** d;   **4.** c;
5. d;   **6.** e;   **7.** e;   **8.** a;   **9.** b.;   **10.** a;   **11.** c;   **12.** c;   **13.** c;
14. c;   **15.** c.

Chapter 5. **1.** organs;   **2.** protection, sensation, thermoregulation, vitamin
synthesis, nutrient storage, excretion;   **3.** cutaneous membrane, accessory
structures;   **4.** epidermis, dermis;   **5.** e, a, c, b, d, e;   **6.** eleidin;   **7.** 14;
8. a. hair shaft   b. nerve   c. sebaceous gland   d. erector muscle   e. sweat duct
f. hair follicle   g. sweat gland   h. blood vessels   i. fat   j. subcutaneous layer
k. dermis l. epidermis;   **9.** 1.5 - 4mm;   **10.** epidermal;   **11.** finger;
12. melanin;   **13.** germinativum;   **14.** melanin;   **15.** cancer;   **16.** papillary,
reticular;   **17.** papilla;   **18.** irregular;   **19.** elastin;   **20.** carcinoma;
21. squamous;   **22.** moles;   **23.** birthmarks;   **24.** follicle;   **25.** matrix;
26. medulla, cortex;   **27.** cuticle;   **28.** vellus, terminal, intermediate;
29. arrector pili;   **30.** melanocytes;   **31.** air;   **32.** hirsutism;
33. sebaceous;   **34.** sebum;   **35.** apocrine;   **36.** eccrine;   **37.** sensible;
38. ceruminous glands;   **39.** bed, root;   **40.** cuticle, eponychium;
41. hyponychium;   **42.** nervous;   **43.** hypodermis;   **44.** adipose;
45. inflammatory;   **46.** cellulitis;   **47.** decubitis;   **48.** g, e, f, d, a, c, b;
49. hyperkeratosis;   **50.** cyst;   **51.** hyperpigmentosis, melanocyte;
52. seborrheic dermatitis;   **53.** dermatitis.   Self test. **1.** b;   **2.** a;   **3.** b;
4. e;   **5.** c;   **6.** d;   **7.** c;   **8.** d;   **9.** a;   **10.** d;   **11.** b;   **12.** b;   **13.** d;
14. e;   **15.** e.

Chapter 6. **1.** specialization, functional independence, dependence on other organ
systems, integration of activity;   **2.** a, b, d, b, c, f, b, g, b, e, f, d, f, f, c, d, e, g, e, j, k,
j, i, g, h, i, j, i;   **3.** epithelium;   **4.** ectoderm, mesoderm, endoderm;
5. standing;   **6.** anterior surface;   **7.** posterior;   **8.** inferior;   **9.** medial;
10. lateral;   **11.** distal;   **12.** proximal;   **13.** rostral;   **14.** see figure 6.11 of
textbook;   **15.** a. cranial b. spinal cavity c. pelvic d. abdominal e. abdominopelvic
f. thoracic;   **16.** transverse;   **17.** sagittal;   **18.** frontal;   **19.** f, a, b, c, e, d;

20. coelom;   21. thoracic, pericardial, abdominopelvic;   22. pleural;
23. mediastinum;   24. diaphragm;   25. pelvic , abdominal;   26. dorsal;
27. cranial, spinal;   28. fascia;   29. framework.

Chapter 7. l. support, protection, blood cell generation, mineral storage, leverage
for movement;   2. calcium phosphate;   3. flexibility;   4. compact, spongy;
5. osteon;   6. blood vessels;   7. canaliculi;   8. lamellae;   9. interstitial
lamellae;   10. a. lamella b. osteocyte c. blood vessel d. Haversian canal
e. canaliculi;   ll. surface;   12. spongy;   13. trabeculae;   14. compact,
spongy;   15. marrow;   16. growth, repair;   17. capsule;   18. endosteum;
19. osteoclasts;   20. dissolve;   21. bone;   22. osteoblasts;   23. connective;
24. dermal;   25. skull, clavicle, lower jaw;   26. chondrocytes;
27. periosteum;   28. osteoblasts;   29. spongy bone;   30. ossification;
31. spongy bone;   32. osteoclasts;   33. perforating;   34. lamellae;
35. osteoclasts;   36. metaphyses;   37. length;   38. epiphyseal;
39. calcitonin, parathyroid hormone;   40. terminates;   41. ossified;   42. 18;
43. electric;   44. formation;   45. larger, heavier;   46. exercise;   47. muscle,
nerve;   48. calcitonin;   49. reduce;   50. parathyroid hormone;
51. increase;   52. See figure 7.8 of textbook;   53. k, l, a, d, i, g, j, e, h, b;
54. hematoma;   55. callus;   56. callus;   57. cartilage;   58. osteoblasts,
osteoclasts;   59. long, short, flat, irregular, wormian, sesamoid;   60. patella;
61. heterotopic;   62. myositis ossificans;   63. process, trochanter, tuberosity,
tubercle, crest, line, head, condyle, trochlea, facet, spine, fossa, sulcus, alveolus,
foramen, fissure, meatus, sinus;   64. articulation;   65. synarthosis, amphiarthrosis,
diarthrosis;   66. sutures;   67. connective;   68. gomphosis;
69. epiphyseal;   70. syndesomosis;   71. symphysis;   72. long;
73. cartilage;   74. capsule;   75. reduce;   76. ligaments;   77. menisci;
78. bursae;   79. dislocation;   80. ankylosis;   8l. rheumatism;
82. osteoarthritis;   83. rheumatoid;   84. gliding, angular, rotational;
85. pivots, hinges;   86. pivot;   87. hinge;   88. gliding, elipsoid, saddle;
89. elipsoid;   90. triaxial;   91. h, m, b, a, c, f, g, d, e, k, l, j, i, n;   92. c, d, e, f,
m, j, k, l, b, a, g, i, h.   Self test. l. e;   2. c;   3. b;   4. a;   5. c;   6. e;
7. d;   8. b;   9. a;   10. c;   ll. a;   12. d;   13. e;   14. d;   15. c.

Chapter 8. l. axial;   2. 80;   3. appendicular;   4. 126;   5. 22, 8, 14;
6. c, a, d, f, e, b;   7. occipital, parietal, frontal, nasal, maxilla, lacrimal, mandible,
zygomatic, temporal, vomer, ethmoid, sphenoid, palatines;   8. frontal, maxilla,
ethmoid, sphenoid, lacrimal, zygomatic;   9. frontal, maxillae, sphenoid;
10. maxillae;   ll. pituitary;   12. alveolar;   13. olfactory;   14. inferior;
15. hyoid;   16. mandible - n, m, temporal - l, k, sphenoid - h, g, g, e, maxilla - j,
frontal - d, occipital - a, b, c;   17. fontanelles;   18. five years;
19. craniostenosis;   20. cleft;   21. a. frontal   b. sphenoid   c. lacrimal   d. nasal
e. zygomatic   f. maxilla   g. mandible   h. occipital   i. temporal   j. parietal;
22. a. parietal   b. frontal   c. coronal suture   d. infraorbital foramen   e. middle nasal
concha   f. mandible   g. mental foramen   h. vomer   i. inferior nasal concha
j. maxilla   k. nasal   l. zygomatic   m. lacrimal   n. ethmoid   o. temporal   p. sphenoid
q. supraorbital foramen   r. sagittal suture   s. occipital   t. lambdoidal suture
u. mastoid process   v. occipitomastoid suture;   23. a. crista galli b. cribiform platae
c. frontal bone d. anterior fossa e. sphenoid bone f. temporal bone g. middle fossa
h. sella turcica i. petrous portion of temporal j. parietal k. posterior fossa l. occipital
bone m. foramen magnum n. internal auditory meatus o. jugular foramen p. foramen
lacerum q. foramen ovale r. foramen rotundum s. optic foramen;   24. 7, 12, 5, 1, 1;

25. 4;   **26.** centrum;   **27.** pedicles;   **28.** lamina;   **29.** transverse;
30. articulating;   **31.** disc;   **32.** shock;   **33.** atlas, axis;   **34.** axis;
35. rib;   **36.** 12;   **37.** true;   **38.** false;   **39.** floating;   **40.** manubrium,
body, xiphoid process;   **41.** lumbar;   **42.** herniated;   **43.** 5;   **44.** hip;
45. spina bifida;   **46.** a. cervical b. thoraci c. lumbar d. sacral;   **47.** cervical -
a. spinous process, b. lamina c. superior articulating surface d. transverse foramen
e. centrum f. pedicle thoracic - a. superior articulating process b. centrum c. rib
facet d. inferior articulating process e. spinous process lumbar - a. superior articular
surface b. spinous process c. inferior articular process d. pedicle e. body
f. transverse process;   **48.** kyphosis;   **49.** scoliosis;   **50.** lordosis.   Self
test. **1.** c;   **2.** c;   **3.** b;   **4.** b;   **5.** b;   **6.** b;   **7.** e;   **8.** b;   **9.** a;
10. b;   **11.** a;   **12.** c;   **13.** b;   **14.** d;   **15.** e.

Chapter 9. **1.** radius, ulna;   **2.** glenoid;   **3.** acromion;   **4.** acromial;
5. separation;   **6.** shoulder;   **7.** labrum;   **8.** loose;   **9.** glenohumeral;
10. coracorohumeral;   **11.** muscles;   **12.** humerus;   **13.** muscles;
14. anatomical;   **15.** surgical;   **16.** deltoid;   **17.** trochlea, capitulum;
18. a. scapular notch b. coracoid process c. acromion d. humerus e. lateral border
f.inferior angle g. glenoid fossa h. medial border i. infraspinous fossa
j. supraspinous fossa k. superior angle l. superior border;   **19.** a. clavicle
b. coracoacromial ligament c. tendon of biceps d. subcoracoid bursa
e. glenohumeral ligaments f. subscapular bursa g. subscapularis muscle h. glenoid
labrum i. glenoid cavity j. articular capsule k. subacromial bursa l. acromion
m. acromioclavicular ligament;   **20.** radius;   **21.** pivot;   **22.** trochlea;
23. olecranon;   **24.** trochlear;   **25.** ulna;   **26.** tuberosity;   **27.** olecranon;
28. flexion;   **29.** interlock;   **30.** 8;   **31.** elipsoid, gliding;   **32.** metacarpal;
33. saddle;   **34.** elipsoid;   **35.** 2;   **36.** elipsoid;   **37.** 27;   **38.** mobility;
39. a. surgical neck b. lateral epicondyle c. olecranon fossa d. trochlea e. medial
epicondyles f. coronoid fossa g. anatomical neck h. greater tubercle i. lesser
tubercle j. intertubercular sulcus k. deltoid tuberosity l. lateral epicondyle
m. capitulum;   **40.** a. olecranon b. trochlear notch c. head d. neck e. radius
f. styloid process g. styloid process h. head i. ulna j. coronoid process
k. tuberosity of radius;   **41.** a. phalanges b. metacarpals c. trapezium d. trapezoid
e. capitate f. scaphoid g. hamate h. triquetrum i. pisiform j. lunate k. ulna l. radius;
42. ilium, ischium, pubis;   **43.** symphysis;   **44.** acetabulum;   **45.** iliac;
46. ischium;   **47.** amphiarthrosis;   **48.** obturator;   **49.** sacrum, coccyx;
50. true;   **51.** femur;   **52.** head;   **53.** trochanters;   **54.** line;   **55.** patellar;
56. gliding;   **57.** tibia, fibula;   **58.** tibia;   **59.** tibia;   **60.** malleolus;
61. malleolus;   **62.** hinge;   **63.** three;   **64.** menisci;   **65.** a. crest of ilium
b. anterior superior spine c. acetabulum d. acetabular notch e. superior ramus of
pubis f. pubic tubercle g. pubic crest h. obturator foramen i. inferior ramus of
pubis j. ramus of ischium k. tuberosity of ischium l. ischial spine m. greater
sciatic notch n. posterior superior spine;   **66.** a. greater trochanter b. gluteal
tuberosity c. linear aspera d. intercondylar notch e. medial condyle f. lesser
trochanter g. neck h. head i. patellar surface j. lateral condyle;
67. a. medial condyle b. tuberosity c. anterior crest d. tibia e. medial malleolus
f. lateral condyle g. fibula h. lateral malleolus;   **68.** 7;   **69.** cruciate;
70. a. lateral condyle b. posterior cruciate c. fibular collateral ligament d. patellar
surface e. medial condyle f. tibial collateral ligament g. medial meniscus
h. anterior cruciate;   **71.** talus;   **72.** calcaneous;   **73.** metatarsal;   **74.** 2;
75. longitudinal, transverse;   **76.** a. first phalanx b. second phalanx c. third phalanx
d. metatarsals e. cuboid f. calcaneus g. talus h. navicular i. cuneiform;   **77.** c, a, e, b,

d;   **78.** gliding diarthrosis, ball and socket, hinge diarthrosis, pivot diathrosis, ellipsoidal diarthrosis, ellipsoidal diarthrosis, hinge diarthrosis, gliding diarthrosis, amphiarthrotic symphysis, ball-and-socket diarthrosis, hinge diarthrosis, gliding diarthrosis, gliding diarthrosis, hinge, ellipsoidal diarthrosis, gliding diarthrosis; **79.** a. rib   b. sacrum   c. coccyx   d. femur   e. fibula   f. calcaneus   g. skull h. cervical vertebra   i. scapula   j. sternum   k. thoracic vertebra   l. humerus m. lumbar vertebra   n. radius   o. ulna   p. carpals   q. metacarpals   r. phalanges s. patella   t. tibia   u. clavicle   v. ilium.   Self test. l. a;   2. c.;   3. c; **4.** b.;   **5.** ellipsoid;   **6.** c;   **7.** c;   **8.** b;   **9.** c;   **10.** b.;   **11.** d.; **12.** b;   **13.** a;   **14.** 2;   **15.** c.

Chapter 10. **1.** contractility, extensibility, elasticity;   **2.** locomotion, posture, supporting soft tissue, guarding entrances and exits, movement along internal passages, regulating blood flow, maintaining body temperature;   **3.** skeletal; **4.** epimysium;   **5.** perimysium;   **6.** fasciculus;   **7.** endomysium; **8.** tendon;   **9.** f, b, l, m, g, e, c, b, d, j, h, i, k;   **10.** 10,000;   **11.** fibers; **12.** striated;   **13.** a. myofibril b. fiber c. endomysium d. fasciculus e. perimysium f. epimysium;   **14.** myosin;   **15.** troponin, tropomyosin;   **16.** active; **17.** inward;   **18.** shorten;   **19.** sliding;   **20.** sodium, potassium;   **21.** positive, negative;   **22.** potential;   **23.** transmembrane;   **24.** volt;   **25.** -85; **26.** depolarization;   **27.** hyperpolarization;   **28.** excitability;   **29.** action; **30.** threshold;   **31.** sodium;   **32.** potassium;   **33.** sodium;   **34.** refractory period;   **35.** See figure 10.6 in the textbook;   **36.** currents;   **37.** all or none; **38.** calcium;   **39.** troponin;   **40.** active;   **41.** cross bridges;   **42.** ADP; **43.** sarcoplasmic;   **44.** tropomyosin;   **45.** motor;   **46.** acetylcholine; **47.** action potential;   **48.** acetylcholinesterase;   **49.** a. neuron b. synaptic cleft c. synaptic vesicles d. ACh receptor site e. muscle cell;   **50.** isotonic; **51.** isometric;   **52.** isometric, isotonic;   **53.** twitch;   **54.** 10 msec, 40 msec, 50 msec;   **55.** length;   **56.** cross bridges;   **57.** a. treppe b. summation c. incomplete tetanus d. complete tetanus;   **58.** tetanus;   **59.** calcium; **60.** motor units;   **61.** tone;   **62.** spasm;   **63.** creatine phosphate;   **64.** CPK; **65.** ATP;   **66.** glycolysis;   **67.** lactic;   **68.** anaerobic glycolysis;   **69.** lactic acid;   **70.** rigor mortis;   **71.** glucose;   **72.** glycogen;   **73.** oxygen debt; **74.** myoglobin;   **75.** slow;   **76.** endurance;   **77.** red;   **78.** slow; **79.** anaerobic capacity;   **80.** hypertrophy;   **81.** lipids, amino acids; **82.** circulatory, respiratory;   **83.** anaerobic;   **84.** aerobic;   **85.** atrophy; **86.** initial, recovery;   **87.** shivering;   **88.** intercalated;   **89.** neural; **90.** refractory;   **91.** tetanus;   **92.** plasticity;   **93.** multi-unit;   **94.** visceral; **95.** f, e, d, b, a, g, c.   Self test. l. e;   2. a;   3. b;   4. c;   5. b;   6. e; 7. c;   8. c;   9. c;   10. c;   11. d   12. b;   13. c;   14. d;   15. c.

Chapter 11. **1.** parallel;   **2.** 30;   **3.** convergent;   **4.** pennate; **5.** parallel;   **6.** deltoid;   **7.** sphincter;   **8.** fulcrum;   **9.** speed;   **10.** force; **11.** force, speed, movement;   **12.** origin, insertion;   **13.** action;   **14.** skeleton; **15.** prime mover;   **16.** synergist;   **17.** antagonistic;   **18.** l, a, b, c, j, k, p, g, i, m, d, n, o, h, f;   **19.** facial;   **20.** a. frontalis b. orbicularis oculi c. nasalis d, zygomaticus e. orbicularis oris f. risorius g. buccinator h. platysma i. occipitalis j. auricular; **21.** masseter, temporalis;   **22.** a. superior rectus b. superior oblique c. medial rectus d. lateral rectus e. inferior rectus f. inferior oblique;   **23.** oculomotor; **24.** abducens;   **25.** trochlear;   **26.** genioglossus;   **27.** palatoglossus; **28.** styloglossus;   **29.** hyoglossus;   **30.** e, h, a, g, f, c, d;   **31.** semispinalis capitis, splenius, spinalis dorsi;   **32.** capitis, cervicis, dorsi;   **33.** transversus;

34. longus capitis, longus cervicis, quadratus lumborum;   **35.** a, e, g, c, f, d, b, h, i;
36. inguinal, hiatal;   **37.** perineum;   **38.** urogenital;   **39.** bulbocavernosus;
40. pudendal;   **41.** bulbocavernosus, ischiocavernosus;   **42.** coccygenal, external
anal sphincter, iliococcygeus, pubococcygeus;   **43.** b, c, d, e, g, f, a;
44. supraspinatus, infraspinatus, subscapularis;   **45.** deltoid;
46. coracobrachialis;   **47.** scapula;   **48.** pectoralis major;   **49.** latissimus
dorsi;   **50.** a. pectoralis minor b. levator scapulae c. rhomboid major d. rhomboid
minor e. serratus anterior f. trapezius;   **51.** supinator, flexor carpi ulnaris, biceps
brachii, extensor carpi radialis, palmaris longus, triceps, brachialis, extensor carpi
ulnaris, extensor carpi radialis, brachoradialis, pronator quadratus;   **52.** adductor
pollicis;   **53.** flexor digitorum;   **54.** abductor pollicis longus;   **55.** extensor
digitorum;   **56.** a. biceps (short head) b. biceps (long head) c. suprinatus
d. infraspinatus e. teres minor f. teres major g. triceps (lateral head) h. triceps (long
head) i. coracobrachialis;   **57.** a. flexor carpi ulnaris b. flexor digitorum c. abductor
pollicis longus d. extensor carpi ulnaris e. supinator f. extensor carpi radialis
g. brachioradialis h. extensor digitorum;   **58.** maximum, medius, minimus;
59. obturators, piriformis;   **60.** pubic;   **61.** adducts;   **62.** femoral;
63. vastus lateralis, biceps femoris, popliteus, rectus femoris, sartorius,
semimembranosus, gracilis;   **64.** tibialis anterior;   **65.** gastrocnemius
66. extensor digitorum longus;   **67.** flexor digitorum longus;   **68.** peroneus
brevis;   **69.** inverts;   **70.** extensor carpi;   **71.** tennis leg;   **72.** a. gluteus
maximus b. biceps femoris c. semitendinosus d. soleus e. gastrocnemius
f. semimembranosus g. gracilis h. adductor magnus i. sartorius j. adductor longus
k. aductor magnus l. tensor fascia lata m. rectus femoris n. vastus lateralis o. peroneus
longus p. tibialis anterior q. extensor digitorum;   **73.** 1. 22, 2. 21, 3. 20, 4. 19, 5. 18,
6. 17, 7. 16, 8. 15, 9. 14, 10. 13, 11. 12, 12. 11, 13. 10, 14. 9, 15. 8, 16. 7, 17. 6, 18. 5, 19. 4, 20. 3,
21. 1, 22. 43, 23. 42, 24. 41, 25. 40, 26. 39, 27. 38, 28. 37, 29. 36, 30. 35, 31. 34, 32. 33,
33. 32, 34. 31, 35. 30, 36. 26, 37. 27, 38. 29, 39. 28, 40. 25, 41. 23, 42. 24, 43. 2;
74. 1. 15, 2. 16, 3. 17, 4. 18, 5. 19 6. 20. 7. 21, 8. 22, 9. 23, 10. 24, 11. 25, 12. 26, 13. 27,
14. 28, 15. 29, 16. 30, 17. 31, 18. 32, 19. 33, 20. 14, 21. 13, 22. 12, 23. 11, 24. 10, 25. 9, 26. 8,
27. 7, 28. 6, 29. 5, 30. 4, 31. 3, 32. 2, 33. 1.   Self test. **l.** e;   **2.** c;   **3.** a;
4. d;   **5.** e;   **6.** e;   **7.** d;   **8.** e;   **9.** c;   **10.** b;   **ll.** a;   **12.** a;   **13.** e;
14. c;   **15.** e.

Chapter 12. **l.** a. sensation b. integration c. coordination d. control;   **2.** central;
3. peripheral;   **4.** neuron;   **5.** a. dendrites b. soma c. axon d. synaptic terminals;
6. neuroglial;   **7.** one half;   **8.** f, b, c, d, e, a, f, a, d, a, a;   **9.** multipolar;
10. bipolar;   **ll.** unipolar;   **12.** anaxonal;   **13.** multipolar;   **14.** sensory;
15. bipolar;   **16.** 80-90;   **17.** bodies;   **18.** Nissl bodies;   **19.** neurofilaments;
20. hillock;   **21.** collaterals;   **22.** telodendria;   **23.** a. initial segment b. axon
c. collateral branch d. telodendria e. synaptic knobs f. myelin sheath g. axonal hillock
h. dendrites;   **24.** synapse;   **25.** neuroeffector;   **26.** postsynaptic;
27. neurotransmitter;   **28.** action potential;   **29.** vesicles;   **30.** delay;
31. presynaptic, postsynaptic;   **32.** a. mitochondrion b. synaptic knob c. presynaptic
membrane d. synaptic cleft e. postsynaptic membrane f. synaptic vesicles
g. endoplasmic reticulum h. neurofilaments;   **33.** -70;   **34.** -60;   **35.** relative;
36. axons;   **37.** -60;   **38.** sensitive;   **39.** facilitated;   **40.** a. 9 b. 7 c. 6 d. 2
e. 4 f. 8 g. 10 h. 5 i. 3 j. 1;   **41.** increases;   **42.** nodes of Ranvier;
43. saltatory;   **44.** A;   **45.** 18;   **46.** c;   **47.** cholinergic;   **48.** calcium;
49. fatigue;   **50.** sodium;   **51.** EPSP;   **52.** summation;   **53.** temporal;
54. spatial;   **55.** IPSP;   **56.** summate;   **57.** norepinephrine;   **58.** serotonin,
dopamine, GABA;   **59.** neuromodulator;   **60.** aerobic;   **61.** 80,000;

62. EPSP;   63. electrical;   64. sensory (afferent);   65. outside;   66. motor
(efferent);   67. interneurons;   68. interneurons;   69. reflexes;
70. monosynaptic;   71. polysynaptic;   72. somatic, visceral;   73. pools;
74. divergence;   75. convergence;   76. reverberating;   77. h, a, f, c, g, d, i, e, b
Self test. l.c;   2. b;   3. b;   4. c;   5. d;   6. c;   7. d;   8. c;   9. c;
10. a;   11. b;   12. d;   13. e;   14. d;   15. e.

Chapter 13. 1. integrating;   2. center;   3. nuclei, higher;
4. exteroceptors;   5. proprioreceptors;   6. interoceptors;   7. d, b, c, g, f, e, h, i,
a;   8. posterior median sulcus;   9. anterior median fissure;   10. enlargements;
11. 31;   12. spinal;   13. sensory;   14. motor;   15. conus medullaris;
16. filum terminale;   17. dura mater;   18. epidural space;   19. arachnoid;
20. CSF;   21. pia mater;   22. denticulate;   23. second;   24. a. posterior
median sulcus b. dorsal root c. ventral root d. spinal nerve e. dorsal root ganglion
f. spinal nerve g. dura mater h. arachnoid mater i. pia mater j. meningers k. rootlets
l. anterior median fissure m. gray matter n. white matter;   25. horns;   26. nuclei;
27. posterior;   28. anterior;   29. lateral;   30. commissure;   31. columns;
32. tracts;   33. ascending, descending;   34. a. posterior median sulcus b. dorsal
root c. somatic d. visceral e. sensory f. motor g. ventral root h. anterior gray commissure
i. anterior median fissure j. anterior gray horn k. lateral gray horn l. posterior gray horn
m. posterior gray commisure n. posterior white column o. anterior white commissure
p. anterior white column q. lateral white column;   35. myelography;   36. 8, 12, 5,
5, 1;   37. T5;   38. epineurium;   39. endoneurium;   40. white;   41. gray;
42. dorsal;   43. ventral;   44. cervical;   45. brachial;   46. lumbar;
47. cervical - c, brachial - a, f, h, lumbar - b, d, e, sacral - g, i, j;   48. a. ventral ramus
b. dorsal ramus c. spinal nerve d. dorsal root ganglion e. white ramus f. autonomic
nerve g. autonomic ganglion h. gray ramus;   49. stretch;   50. inhibited;
51. interconnecting;   52. a. receptor b. sensory neuron c. excitatory d. inhibitory
e. inhibitory f. excitatory g. excitatory;   53. centers;   54. c, a, b, f, d, e.   Self
test. 1. c;   2. c;   3. b;   4. b;   5. d;   6. d;   7. a;   8. b;   9. b;
10. b;   11. a;   12. b;   13. c;   14. e;   15. b.

Chapter 14. 1. cortex;   2. cranial;   3. telencephalon, diencephalon,
mesencephalon, metencephalon, myelencephalon;   4. dural sinuses;   5. falx
cerebri;   6. tentorium cerebelli;   7. choroid plexuses;   8. diffusion;
9. cerebrum;   10. thalamus;   11. cerebral aqueduct;   12. 4th;   13. arachnoid;
14. a. lateral ventricle b. cerebral hemisphere c. interventricular foramen d. third
ventricle e. pons f. medulla g. fourth ventricle h. cerebellum i. cerebral aqueduct;
15. see figure 14.3 in the textbook;   16. gyri;   17. sulcus;   18. longitudinal;
19. frontal, parietal, temporal, insula, occipital;   20. central sulcus;
21. precentral;   22. postcentral;   23. association;   24. association;
25. commissural;   26. projection;   27. anterior commissure, corpus callosum;
28. internal;   29. e, d, i, a, f, b. c, g, h;   30. a. cingulate gyrus b. mamillary body
c. parahippocampal gyrus d. hippocampus e. temporal lobe f. fornix g. corpus callosum
h. central sulcus;   31. pineal gland, thalamus, hypothalamus;   32. endocrine;
33. sensory;   34. c, a, b, e, d;   35. mamillary bodies, preoptic area, supraoptic
nucleus, autonomic centers, paraventricular nucleus;   36. endocrine;   37. corpora
quadrigemina;   38. visual;   39. auditory;   40. red;   41. substantia;
42. cerebral peduncles;   43. pons, cerebellum;   44. vermis;   45. postural;
46. motor;   47. pons;   48. medulla;   49. tracts;   50. medulla oblongata;
51. medulla;   52. olivary;   53. five;   54. reticular;   55. a. anterior
commissure b. hypothalamus c. hypophysis d. temporal lobe e. pons f. spinal cord
g. medulla oblongata h. cerebellum i. pineal body j. occipital lobe k. thalamus l. parietal

lobe m. fornix n. corpus callosum o. frontal lobe;   **56.** vagus, hypoglossal, olfactory, abducens, trochlear, oculomotor, spinal accessory, acoustic, trigeminal, glossopharyngeal, facial;   **57.** ophthalmic, maxillary, madibular;   **58.** cochlear, vestibular;   **59.** semilunar;   **60.** geniculate;   **61.** glossopharyngeal; **62.** motor - oculomotor, trochlear, abducens, spinal accessory, hypoglossal: sensory - olfactory, acoustic, optic: mixed - trigeminal, facial, glossopharyngeal, vagus,; **63.** a. olfactory b. optic c. oculomotor, d. trochlear e. trigeminal f. abducens g. facial h. acoustic i. glossopharyngeal j. vagus, k. spinal accessory l. hypoglossal; **64.** migrain;   **65.** concussion;   **66.** blood;   **67.** encephalitis Self test. **1.** c,; **2.** c;   **3.** b;   **4.** c;   **5.** b;   **6.** e;   **7.** d;   **8.** d;   **9.** a;   **10.** e;   **11.** b; **12.** c;   **13.** d;   **14.** a;   **15.** a

Chapter 15. **1.** processing;   **2.** neurotransmitters;   **3.** 30 -50; **4.** norepinephrine, serotonin;   **5.** dopamine;   **6.** enkephalin, endorphin; **7.** medulla, reticular formation, thalamus;   **8.** motor;   **9.** anencephaly; **10.** brainstem;   **11.** nuclei;   **12.** 1. h 2. k 3. d 4. g 5. 1 6. c 7. f 8. i 9. e 10. j 11. a 12. b; **13.** spino;   **14.** spino;   **15.** fasciculus cuneatus, fasciculus gracilis; **16.** medulla;   **17.** spinothalamic tract;   **18.** right;   **19.** thalamus; **20.** spinocerebellar;   **21.** pyramidal;   **22.** corticobulbar;   **23.** lateral; **24.** commissure;   **25.** left;   **26.** rubrospinal, vestibulospinal, reticulospinal, tectospinal;   **27.** rubrospinal;   **28.** vestibulospinal;   **29.** rubrospinal, vestibulospinal;   **30.** tectospinal;   **31.** cerebral;   **32.** initiate;   **33.** caudate, putamen;   **34.** spinocerebellar;   **35.** vestibular;   **36.** voluntary, involuntary; **37.** predicting;   **38.** pyramidal, extrapyramidal, cerebellum;   **39.** dopamine; **40.** delta, alpha, beta, theta;   **41.** REM;   **42.** slow wave;   **43.** 5;   **44.** slow wave;   **45.** central nervous system;   **46.** RAS;   **47.** norepinephrine, serotonin; **48.** short term;   **49.** tertiary memories;   **50.** reverberating;   **51.** engram; **52.** limbic system;   **53.** learning;   **54.** retrograde;   **55.** anterograde; **56.** interpretive;   **57.** prefrontal cortex;   **58.** dominant;   **59.** nondominant; **60.** left;   **61.** dominant;   **62.** e, c, f, b, d, a, Self test. **1.** d;   **2.** d;   **3.** e; **4.** c;   **5.** b;   **6.** d;   **7.** b;   **8.** e;   **9.** c;   **10.** e;   **11.** a;   **12.** d; **13.** d;   **14.** c;   **15.** d

Chapter 16. **1.** visceral;   **2.** synapse;   **3.** first order;   **4.** second order; **5.** preganglionic, postganglionic;   **6.** thoracolumbar;   **7.** craniosacral;   **8.** L2; **9.** white;   **10.** paravertebral;   **11.** gray;   **12.** autonomic;   **13.** 1:32;   **14.** a. preganglionic fiber b. spinal cord c. autonomic (sphanchnic nerve) d. sympathetic ganglion e. postganglionic fiber f. gray ramus g. white ramus h. spinal nerve i. posterior root ganglion j. visceral sensory neuron;   **15.** paravertebral (chain);   **16.** splanchnic;   **17.** celiac ganglion, superior mesenteric ganglion;   **18.** celiac ganglion;   **19.** superior mesenteric ganglion;   **20.** inferior mesenteric ganglion; **21.** inferior mesenteric ganglion;   **22.** short, long;   **23.** long, short;   **24.** divergence;   **25.** third, seventh, ninth, tenth;   **26.** ten;   **27.** cardiac, pulmonary, hypogastric, celiac;   **28.** acetylcholine;   **29.** acetylcholine, norepinephrine; **30.** alpha, beta;   **31.** alpha;   **32.** depolarization;   **33.** epinephrine;   **34.** second;   **35.** acetylcholine;   **36.** MAO, COMT;   **37.** nicotinic;   **38.** muscarinic;   **39.** alpha;   **40.** beta blockers;   **41.** parasympathetic;   **42.** sympathetic;   **43.** p, p, p, s, s, p, s, s, s, s, p,;   **44.** visceral;   **45.** tone;   **46.** one half;   **47.** centers;   **48.** hypothalamus;   **49.** biofeedback Self Test. **1.** b; **2.** c;   **3.** a;   **4.** d;   **5.** c;   **6.** c;   **7.** a;   **8.** c;   **9.** b;   **10.** e;   **11.** b; **12.** d;   **13.** a;   **14.** e;   **15.** b

Chapter 17. **1.** nerve; **2.** encapsulated; **3.** accessory; **4.** modified;
5. general, special; **6.** general; **7.** exteroeceptors, enteroceptors,
proprioreceptors; **8.** mechanoreceptors, chemoreceptors, photoreceptors,
nociceptors, thermoreceptors; **9.** b - c- a; **10.** labeled; **11.** modality;
12. field; **13.** frequency; **14.** tonic, phasic; **15.** perception; **16.** nuclei;
17. peripheral; **18.** central; **19.** free; **20.** Class A, Class C; **21.** referred;
22. hydrogen, oxygen; **23.** skin; **24.** free; **25.** changing; **26.** tactile,
baroreceptors, proprioreceptors; **27.** f, a, d, e, c, b; **28.** pressure;
29. elastic; **30.** tendon; **31.** muscle spindle; **32.** olfactory;
33. olfactory; **34.** mucous; **35.** chemoreceptors; **36.** 4; **37.** bulbs;
38. olfactory; **39.** limbic, hypothalamus; **40.** buds; **41.** taste hair;
42. papillae; **43.** filiform, fungiform, circumvallate; **44.** sour, bitter, sweet,
salt; **45.** sour, bitter, sour, salt; **46.** 7th, 9th, 10th; **47.** solitarius;
48. lemniscus; **49.** olfactory; **50.** a, primary sensory cortex, b. thalamic nucleus,
c. medial lemniscus d. solitary nucleus e. X f. IX, g. VII h. sweet i. salt j. sour
k. bitter; **51.** m, s, h, b, a, g, n, k, r, o, i, p, e, f, d, a, i, q; **52.** otitis media;
53. auditory ossicles; **54.** a. tympanic membrane b. middle ear c. inner ear
d. acoustic nerve e. semicircular canal f. cochlea g. pharyngotympanic tube h. vestibule
i. external auditory meatus j. auditory canal; **55.** hair; **56.** stereocilia;
57. receptor; **58.** lateral, superior, posterior; **59.** ampulla; **60.** cupula;
61. endolymph; **62.** utricle; **63.** endolymphatic sac; **64.** acceleration;
65. macula; **66.** otoconia; **67.** otoconia; **68.** vestibular; **69.** vestibular;
70. vertigo; **71.** tympanic membrane; **72.** ossicles; **73.** malleus, incus;
74. stapes; **75.** 22; **76.** a. endolymphatic sac b. endolymphatic duct
c. vestibular branch d. vestibular ganglia e. cochlear branch f. cochlear duct
g. saccule h. utricle i. lateral semicricular canal j. ampulla k. posterior
semicircular canal l. superior semicircular canal; **77.** scala vestibuli, scala media,
scala tympani; **78.** vestibuli, tympani; **79.** round window; **80.** Corti;
81. basilar; **82.** tectorial; **83.** tectorial; **84.** receptor; **85.** base (round
window); **86.** conduction; **87.** nerve; **88.** spiral; **89.** cochlear;
90. inferior colliculus, auditory cortex; **91.** organ of Corti; **92.** direction;
93. a. cochlear duct b. organ of Corti c. basilar membrane d. scala tympani
e. spiral ganglion f. scala vestibuli g. tectorial membrane; **94.** canthus, canthus;
95. conjunctiva; **96.** lacrimal; **97.** a. lacrimal canals b. lacrimal puncta,
c. nasolacrimal duct d. lateral canthus e. inferior lacrimal gland f. superior
lacrimal gland; **98.** q, k, a, e, m, o, n, b, g, l, t, f, p, j, i, s, r, h, b; **99.** a. pupil
b. conjunctiva c. posterior chamber d. ciliary muscle e. suspensory ligament
f. lateral rectus, g. vitreous body h. fovea centralis i. optic nerve j. sclera
k. choroid l. retina m. medial rectus n. iris o. anterior chamber p. cornea;
100. accommodation; **101.** flatter; **102.** contracts; **103.** elasticity;
104. further; **105.** acuity; **106.** 20; **107.** 200; **108.** photon; **109.** rod,
cone; **110.** night, day; **111** . red, blue, green; **112.** rhodopsin; **113.** retinene;
114. breakdown; **115.** rods, cones; **116.** night; **117.** fovea centralis; **118.** red;
119. ganglion, bipolar neuron; **120.** acuity; **121.** bipolar; **122.** blind;
123. chiasma; **124.** geniculate; **125.** visual; **126.** colliculus;
127. a. photoreceptor layer b. cone cell c. rod cell d. bipolar neuron layer
e. ganglion cell layer f. optic nerve Self test. **1.** b; **2.** e; **3.** d; **4.** e;
5. a; **6.** e; **7.** b; **8.** e; **9.** c; **10.** e; **11.** b; **12.** d; **13.** b;
14. a; **15.** c

Chapter 18. **1.** hormones; **2.** amino acid, polypeptides, steroids;
3. epinephrine, norepinephrine; **4.** polypeptides; **5.** steroid; **6.** a. hypophysis

b. thyroid c. parathyroid d. pancreas e. ovary f. testes g. adrenal h. pineal gland;
7. enzymes;   **8.** receptor;   **9.** genes;   **10.** membrane;   **11.** second;
12. cyclic AMP;   **13.** adenyl cyclase, activation;   **14.** cytoplasm;
15. transcription;   **16.** gene, enzyme;   **17.** nucleus;   **18.** genes;   **19.** see table
18.1 in textbook;   **20.** adrenal medulla;   **21.** endocrine;   **22.** releasing,
inhibiting;   **23.** a. infundibulum   b. neurohypophysis   c. hypophysis
d. hypothalamus   e. pars intermedia   f. adenohypophysis;   **24.** hypophysis;
25. neurohypophysis;   **26.** supraopticl, paraventricular;   **27.** water ;
28. oxytocin;   **29.** axoplasmic;   **30.** adenohypophysis;   **31.** f, a, g, d, b, e, c;
32. hypophyseal portal system;   **33.** releasing;   **34.** inhibiting;
35. inhibiting;   **36.** isthmus;   **37.** a. hyoid b. left lateral lobe c. follicles d. trachea
e. isthmus f. right lateral lobe g. thyroid cartilage;   **38.** follicles;   **39.** tyrosine;
40. thyroxine;   **41.** metabolism;   **42.** thyroglobulin;   **43.** thyroid binding
globulin (TBG);   **44.** calcitonin;   **45.** decrease;   **46.** chief;   **47.** increase;
48. p, p, c, c, c;   **49.** thymosin;   **50.** cortex, medulla;   **51.** reticularis,
fasciculata, glomerulosa;   **52.** male;   **53.** glucocorticoids;   **54.** glucose,
glycogen;   **55.** cortisol, cortisone, corticosterone;   **56.** inflammatory;
57. mineralocorticoids;   **58.** aldosterone;   **59.** sodium;   **60.** angiotensin II;
61. catecholamines;   **62.** sympathetic;   **63.** first;   **64.** i, i, i, d, d,;
65. a. adrenal gland, b. cortex c. capsule d. medulla e. kidney;   **66.** renin;
67. erythropoietin;   **68.** depresses thirst, promotes sodium loss, blocks secretion of
water conserving hormones, dilates peripheral circulation;   **69.** glucagon, insulin;
70. insulin;   **71.** lipids, glycogen;   **72.** I;   **73.** increase;   **74.** glucose;
75. a. alpha cells,   b. beta cells,   c. acinar cells   d. blood vessel;
76. testosterone;   **77.** secondary;   **78.** inhibin;   **79.** estrogen;   **80.** corpus
luteum;   **81.** development;   **82.** progesterone;   **83.** anterior pituitary;
84. melatonin;   **85.** MSH;   **86.** gonadotrophin releasing factor;
87. synergistic;   **88.** GH, thyroid hormones, insulin, PTH, gonadal hormones;
89. stress;   **90.** GAS;   **91.** catecholamines;   **92.** glucocorticoids;
93. mobilization of lipid and protein reserves, elevation and stabilization of blood
glucose, conservation of glucose for neural tissues;   **94.** exhaustion;
95. behavior;   **96.** i, e, f, h, d, c, a, b, g   Self test. **1.** b;   **2.** c;   **3.** b;   **4.** c;
5. c;   **6.** d;   **7.** c;   **8.** c;   **9.** c;   **10.** a;   **11.** b;   **12.** d;   **13.** c;   **14.** b;
15. e

Chapter 19. **1.** nutrients, gases, wastes, hormones;   **2.** temperature regulation,
defense, pH balance, prevention of fluid loss;   **3.** plasma;   **4.** formed (cellular);
5. 5-6, 4-5;   **6.** hypervolemic;   **7.** 55;   **8.** water;   **9.** proteins, oxygen;
10. albumins, globular proteins, fibrinogen;   **11.** albumins;   **12.** immunoglobulins;
13. metalloproteins;   **14.** urine;   **15.** lipoproteins;   **16.** fibrinogen;   **17.** red
cells, white cells, platelets;   **18.** hematocrit;   **19.** 40 -54, 37 - 47;   **20.** 5.2;
21. thick, thin;   **22.** surface;   **23.** rouleaux;   **24.** nuclei;   **25.** 120;   **26.** 3
million;   **27.** hemoglobin;   **28.** protein, heme;   **29.** iron;   **30.** oxygen;
31. bilirubin;   **32.** transferrin;   **33.** ferritin, hemosiderin;   **34.** see figure 19.5 in
textbook;   **35.** exposed;   **36.** AB;   **37.** 0;   **38.** negative;   **39.** O
negative;   **40.** negative, positive;   **41.** granulocytes, agranulocytes;
42. immune;   **43.** peripheral;   **44.** 6000;   **45.** d, a, a, e, b, c, b, a, e;
46. megakaryocyte;   **47.** enzymes;   **48.** see figure 19.2 and 19.8 in your textbook;
49. hemostasis;   **50.** vascular;   **51.** platelet;   **52.** coagulation;   **53.** fibrin;
54. proenzymes;   **55.** extrinsic;   **56.** a. calcium b. prothrombin c. fibrinogen;
57. retraction;   **58.** plasmogen;   **59.** embolus;   **60.** heparin, coumadin,
EDTA;   **61.** streptokinase, urokinase;   **62.** hemopoiesis;   **63.** myeloid;

64. oxygen;   65. hemocytoblast;   66. erythroblast;   67. reticulocyte;
68. granulocyte;   69. lymphocytes;   70. g, i, a, d, c, f, h, b, e   Self test. l. e;
2. b;   3. b;   4. d;   5. d;   6. c;   7. d;   8. e;   9. d;   10. a;   11. d;
12. c;   13. b;   14. d;   15. c

Chapter 20. 1. arteries, veins, capillaries;   2. pulmonary, systemic;
3. systemic, pulmonary;   4. atrium, ventricle;   5. mediastinum;
6. pericardial;   7. visceral, parietal;   8. serous;   9. lubricant;   10. a. fibrous
layer, b. serous layer c. pericardial space d. visceral pericardium e. parietal pericardium
f. myocardium;   11. coronary;   12. interventricular;   13. interatrial,
interventricular;   14. j, c, d, j, e, h, i, g, b, f,;   15. a. aorta   b. left pulmonary artery
c. left pulmonary veins   d. left atrium   e. bicuspid valve   f. aortic semilunar valve
g. left ventricle   h. right ventricle   i. papillary muscle   j. inferior vena cava
k. chordae tendinae   l. tricuspid valve   m. right atrium   n. pulmonary semilunar
valve   o. right pulmonary veins   p. right pulmonary artery   q. superior vena cava;
16. coronary;   17. marginal, posterior interventricular;   18. circumflex,
interventricular;   19. ischemia;   20. angina pectoris;   21. myocardial;
22. coronary;   23. epicardium, myocardium, endocardium;   24. intercalated;
25. 300;   26. tetanus;   27. pacemaker;   28. SA node, AV node left and right
bundle branches;   29. interventricular;   30. AV node;   31. atria, ventricles;
32. d, c, a, b,;   33. a. purkinje fibers   b. left-bundle branch   c. right-bundle branch
d. bundle of His   e. atrioventricular node   f. sinoatrial node;   34. t, t, f, t, f, t, f, t,
t;   35. chordae tendinae;   36. AV;   37. ventricular;   38. two;
39. a. aorta,   b. tricuspid valve   c. bicuspid valve   d. pulmonary trunk
e. pulmonary semilunar valve   f. aortic semilunar valve;   40. ECG;   41. P;
42. ventricles;   43. T;   44. a. P b. QRS c. T;   45. stroke volume;
46. cardiac reserve;   47. autonomic;   48. increases;   49. decreases;
50. increase;   51. slow;   52. atrial;   53. ESV;   54. 70;   55. a. venous
return b. diastole c. heart rate d. increase e. Starling's law;   56. increase;
57. force;   58. autonomic;   59. 180;   60. cardiac;   61. cardioacceleratory;
62. cardioinhibitory;   63. baroreceptors;   64. b, a, c, d, e.   Self test. l. d;
2. c;   3. a;   4. c;   5. c;   6. d;   7. b;   8. c;   9. a;   10. c;   11. a;
12. b;   13. d;   14. c;   15. b.

Chapter 21. 1. tunica interna, tunica media, tunica externa;   2. endothelial;
3. tunica media;   4. tunica externa;   5. a. tunica media   b. tunica interna
c. tunica externa   d. tunica interna   e. endothelium   f. elastic membrane   g. tunica
externa   h. tunica media;   6. elastic;   7. elastic;   8. muscular;
9. arterioles;   10. diameter;   11. vasa vasorum;   12. endothelium;   13. 8;
14. fenestrated;   15. sinusoids; 16. capillary;   17. precapillary sphincter;
18. vasomotion;   19. collaterals;   20. metarteriole;   21. externa;
22. valves;   23. venous;   24. reservoir;   25. hemorrhoids;
26. A. a. external iliac   b. femora   c. tibial   d. femoral   e. common iliac
B. a. superior mesenteric   b. hepatic portal   c. hepatic   C. a. brachiocephalic
b. internal carotid   c. posterior cerebral artery   d. brachiocephalic   D. a. subclavian
b. brachial   c. brachial   d. axillary   e. brachiocephalic;   27. right subclavian,
right common carotid;   28. basilar;   29. vertebral;   30. superior mesenteric
vein;   31. celiac;   32. external carotid;   33. brachiocephalic;   34. basilic;
35. lesser saphenous;   36. azygous;   37. cephalic;   38. sinus;   39. a. left
common carotid   b. left subclavian   c. pulmonary   d. axillary   e. brachia   f. renal
g. radial   h. ulnar   i. femoral   j. anterior tibial   k. posterior tibial   l. popiteal
m. external iliac   n. internal iliac   o. common iliac   p. superior mesenteric

q. celiac   r. arch of aorta   s. brachiocephalic   t. right common carotid   u. external
carotid   v. internal carotid;   40. a. subclavian   b. axillary   c. brachial   d. radial
e. basilic   f. ulnar   g. anterior tibial   h. posterior tibial   i. popiteal   j. small
saphenous   k. great saphenous   l. femoral   m. internal iliac   n. external iliac
o. common iliac   p. renal   q. hepatic   r. inferior vena cava   s. superior vena cava
t. right brachiocephalic   u. internal jugular   v. external jugular;   41. a. right
gastric vein   b. gastroepiploic vein   c. splenic   d. inferior mesenteric vein
e. superior mesenteric vein   f. portal vein   g. pyloric vein   h. right and left hepatic
veins   i. inferior vena cava;   42. foramen ovale, ductus arteriosus;   43. internal
iliac;   44. umbilical;   45. ovale;   46. pressure;   47. resistance;
48. viscosity;   49. pressure;   50. circulatory;   51. a. systolic   b. diastolic
c. pulse   d. mean e. elastic;   52. 30;   53. venous return;   54. a. pressure
b. valves   c. skeletal   d. thoracoabdominal;   55. arteriole;   56. dilate,
constrict;   57. vasomotor;   58. vasoconstriction;   59. cardiac output, arterial,
peripheral resistance;   60. local, autonomic, hormonal;   61. venous;
62. increase;   63. low, high, low;   64. decrease;   65. vasodilation;
66. vasoconstriction;   67. viscosity, increase;   68. increase;   69. cardiac
output, peripheral resistance, blood volume;   70. I, D, D, D, D, I, D, D,;   71. carotid
body, aortic body;   72. aldosterone, erythropoietin, ADH, ANF;   73. capillaries;
74. osmotic;   75. dynamic center;   76. hydrostatic;   77. lymphatic;
78. increased;   79. edema;   80. cardiac output;   81. decreased, increased;
82. heart rate;   83. venous;   84. central ischemic;   85. T, F, F, T, T, F, T, T, T,
T, F, T, T, Self test. l. b;   2. d;   3. c;   4. d;   5. c;   6. a;   7. c;
8. e;   9. d;   10. d;   11. e;   12. e;   13. e;   14. b;   15. e

Chapter 22. l. lymph;   2. returns tissue fluid to the circulatory system,
maintains composition of tissue fluid, provides immunity;   3. closed;   4. thoracic,
right lymphatic;   5. lymphedema;   6. NK, T cells, B cells;   7. T cells;
8. cytotoxic;   9. regulatory;   10. B cells;   11. NK;   12. 4;
13. lymphopoiesis;   14. two;   15. NK, B cells;   16. thymosin;   17. lymphatic
nodule;   18. epithelium;   19. tonsils;   20. Peyer's;   21. appendix;
22. thymus, lymph nodes, spleen;   23. thymus;   24. connective tissue;
25. trabeculae;   26. T cells, B cells;   27. a. afferent lymphatic vessel
b. trabecula   c. capsule   d. medullary cord   e. efferent lymphatic vessel   f. hilus
g. medulla   h. cortex;   28. macrophages;   29. lymphoadenopathy;   30. spleen;
31. phagocytosis, immune;   32. white;   33. immune;   34. nonspecific, specific
immunity;   35. epithelium;   36. phagocytic;   37. microphages, macrophages;
38. neutrophil, eosinophil;   39. macrophages;   40. connective;   41. microglia,
Kupfer cells, Langerhans;   42. mobile macrophages;   43. monocytes;
44. diapedesis;   45. immunological;   46. antigens;   47. ll;
48. a. destruction of bacterial cell walls   b. inactivation of viruses   c. stimulation of
inflammation   d. attraction of phagocytes   e. enhancement of phagocytic activity;
49. antibody;   50. properdin;   51. inflammation;   52. antigens;
53. humoral;   54. antigenic determinant;   55. two;   56. haptens;
57. plasma, memory cells;   58. antibody;   59. titre;   60. memory cells;
61. primary;   62. 4;   63. variable;   64. variable;   65. a. complement
b. opsonization   c. attraction of phagocytes   d. stimulation of inflammation
e. prevention of bacterial adhesion;   66. IgD, IgG, IgM, IgE, IgA;   67. T;
68. cytotoxic;   69. antigen;   70. HLA;   71. Class I, Class II;   72. HLA,
HLA;   73. a. accessory   b. HLA   c. T   d. cytotoxic   e. regulatory
f. suppressor   g. regulatory   h. helper i. suppressor;   74. monokines,
lymphokines;   75. interleukin;   76. a. stimulate inflammation   b. formation of

scar tissue   c. induction of fever   d. stimulate Mast cell formation   e. stimulate
ACTH secretion;   77. interferon;   78. BCDF, BCGF;   79. TNF;   80. 4, 2, 1,
3;   81. mother;   82. tolerance;   83. suppression;   84. i, h, e, b, j, d, c, a, g
Self test. 1. b;   2. c;   3. e;   4. d;   5. a;   6. d;   7. c;   8. c;   9. c;
10. c;   11. b;   12. a;   13. d;   14. c;   15. d

Chapter 23. 1. a. provide exchange surface   b. move air   c. protection of
respiratory surface   d. defense of respiratory surface   e. communication;
2. upper, lower;   3. nares;   4. vestibule;   5. septum;   6. hard;   7. soft
palate;   8. nares;   9. conchae;   10. removal;   11. pseudostratified;
12. swallowed;   13. mucosa;   14. humidifies;   15. pharynx;   16. nasopharynx,
oropharynx, laryngopharynx;   17. stratified squamous;   18. c, g, a, b, j, i, d, k, e, h,
f,;   19. air;   20. tension;   21. intrinsic;   22. extrinsic;   23. spasms;
24. laryngitis;   25. trachea;   26. mucosa;   27. cartilage;   28. primary;
29. 3, 2;   30. a. superior concha   b. sphenoid air sinus   c. soft palate   d. pharynx
e. epiglottis   f. larynx   g. esophagus   h. left bronchus   i. left lung   j. right lung
k. right bronchus   l. trachea   m. hard palate   n. inferior concha   o. middle concha
p. frontal air sinus;   31. elastic;   32. a. secondary   b. bronchioles   c. terminal
bronchiole   d. cuboidal e. respiratory;   33. aveolar, alveoli;   34. a.. bronchiole
b. pulmonary arteriole   c. respiratory bronchiole   d. aveolar sac   e. alveolus
f. alveolar duct   g. alveolar capillary network   h. pulmonary venule;
35. squamous;   36. surfactant;   37. macrophages;   38. connective tissue;
39. mucous;   40. pleural;   41. mediastinum;   42. parietal, visceral;
43. potential;   44. pleurisy;   45. collapsed;   46. pulmonary;   47. pulmonary
ventilation, external respiration, internal respiration, cellular respiration;
48. pulmonary ventilation;   49. external;   50. internal;   51. cellular;
52. compressible;   53. increase, decrease;   54. increase;   55. higher, lower;
56. less;   57. a. pleural   b. drop   c. intrapulmonary   d. atmospheric e. decreasing
f. elastic   g. intrapulmonary;   58. 80 mm Hg, 100 mm Hg;   59. pneumothorax;
60. external;   61. internal;   62. elastic rebound;   63. eupnea, hyperpnea;
64. circulation;   65. g,a, d, e, h, f, b, c,;   66. solution;   67. solubility;
68. Henry's;   69. pressure;   70. Dalton's law;   71. partial;   72. 100;
73. dead;   74. partial pressure;   75. higher;   76. partial;   77. higher;
78. 3%;   79. hemoglobin;   80. oxygen-hemoglobin;   81. oxygen;
82. higher;   83. hydrogen ion;   84. release;   85. DPG;   86. higher;
87. higher, 50;   88. plasma, hemoglobin, bicarbonate;   89. 7;
90. carbaminohemoglobin;   91. 70;   92. carbonic anhydrase;
93. bicarbonate;   94. sodium;   95. chloride;   96. a. reticular   b. respiratory
rhythmicity center   c. inspiratory   d. expiration   e. inspiratory center
f. pneumotaxic   g. depth;   97. inspiratory;   98. deflation;
99. Hering-Breuer;   100. forced;   101. aortic, carotid;   102. hydrogen;
103. carbon dioxide;   104. hyperventilation;   105. oxygen;   106. SISD;
107. f, i, b, c, e, g, d, j, a, h Self test. 1. d;   2. b;   3. e;   4. c;   5. c;   6. c;
7. a;   8. e;   9. b;   10. d;   11. a;   12. b;   13. e;   14. c;   15. c

Chapter 24. 1. a. ingestion   b. mechanical processing   c. chemical digestion
d. absorption   e. compaction   f. defecation;   2. epithelium, lamina propria,
muscularis mucosa, submucosa, muscularis externa, adventitia or serosa;
3. squamous columnar;   4. lamina propria, mucosa;   5. mucosa;
6. submucosa;   7. mycnteric;   8. adventitia, serosa;   9. visceral;
10. mesenteries;   11. analysis, mechanical processing, lubrication, digestion;   12. a,
h, b, c, f, g, d, e;   13. mechanical processing, manipulation, sensory analysis;

14. root;   15. lingual;   16. 1.0 to 1.5;   17. lubricate;   18. taste buds;
19. bacterial;   20. alpha-amylase;   21. alpha-amylase;   22. mucin;
23. parasympathetic;   24. salivary;   25. higher;   26. mastication, dentin, pulp
cavity, root canal, alveolus, peridontal ligament, cementum, crown, canines, molars,
deciduous, wisdom, premolars, plaque, peridontal disease;   27. bolus;   28. a. hard
palate,   b. uvula,   c. molars   d. premolars,   e. canine,   f. incisors   g. palatine
tonsil   h. cheeks   i. oropharynx   j. lips;   29. a. epiglottis,   b. palatine tonsil,
c. lingual tonsil   d. circumvallate papillae   e. filiform papillae   f. fungiform
papillae;   30. constrictors;   31. skeletal, smooth;   32. peristalsis;
33. buccal, pharyngeal, esophageal;   34. hard palate;   35. glottis;
36. esophageal;   37. esophageal;   38. reflexive;   39. myenteric;
40. storage, mechanical breakdown, digestion;   41. gastric;   42. a. greater,
b. lesser c. fundus d. body e. pylorus f. pyloric sphincter;   43. rugae;
44. omentum;   45. omentum;   46. mucus;   47. gastric;   48. parietal, chief;
49. pepsin;   50. protein;   51. gastrin;   52. cephalic, gastric, intestinal;
53. cephalic;   54. gastric;   55. muscularis;   56. intestinal;   57. a. fundus
b. serosa   c. circular layer of muscle   d. oblique muscle layer   e. greater curvature
f. rugae   g. omentum   h. pyloric sphincter   i. duodenum   j. longitudinal muscle
layer   k. lower esophageal sphincter   l. diaphragm m. esophagus;   58. liver;
59. 200, a. metabolic regulation, hematological regulation, synthesis and excretion;
60. bile salts;   61. urobilogen;   62. bile salts;   63. enterohepatic;   64. f, a, c,
d, e, b;   65. celiac;   66. lobule;   67. vein;   68. sinusoids;   69. portal;
70. central vein, bile;   71. hepatic;   72. cholecystokinin;   73. emulsifying;
74. a. left lobe,   b. gallbladder,   c. right lobe   d. inferior vena cava   e. falciform
ligament   f. cystic duct   g. right lobe   h. gallbladder   i. quadrate lobe   j. left lobe
k. caudate lobe   l. falciform ligament   m. inferior vena cava;   75. a. central vein
b. cell plates   c. branch of portal vein   d. branch of hepatic artery   e. bile canaliculi
f. sinusoid;   76. alkaline fluid;   77. common bile;   78. secretin,
cholecystokinin;   79. proteinases;   80. proenzymes;   81. 6;   82. duodenum,
jejunum, ileum;   83. plicae;   84. villi;   85. microvilli;   86. lacteal;
87. crypts;   88. bicarbonate;   89. segmentation;   90. submucosal, myenteric;
91. 1.8;   92. dissolves;   93. enterocrinin;   94. gastrin, cholecystokinin, secretin,
GIP, enterocrinin;   95. a. absorption of water and compaction of feces
b. absorption of vitamins   c. storage and elimination of fecal wastes;
96. ileo-cecal;   97. appendix;   98. ascending, transverse, descending, sigmoid,
rectum;   99. goblet;   100. smooth, skeletal;   101. mass;   102. defecation;
103. antigenic;   104. hydrolysis;   105. mouth;   106. small intestines;
107. bacteria;   108. flatus;   109. triglycerides;   110. chylomicrons;
111. lacteal;   112. 150;   113. osmosis;   114. water;   115. small intestines;
116. i, b, c, e, g, k, a, d, j, l, f,;   117. water;   118. a. parotid gland   b. pharynx
c. esophagus   d. stomach   e. pancreas   f. transverse colon   g. jejunum
h. descending colon   i. sigmoid colon   j. rectum   k. appendix   l. cecum
m. ileum   n. ascending colon   o. duodenum   p. gallbladder   q. liver
r. submandibular and sublingual glands   s. mouth;   119. j, g, i, e, f, a, b, c, d, h   Self
test. 1. c;   2. e;   3. c;   4. e;   5. d;   6. b;   7. c;   8. e;   9. b;
10. c;   11. d;   12. b;   13. b;   14. d;   15. c

Chapter 25. 1. nutrients;   2. digestive tract, respiratory tract, circulatory
system;   3. endocrine;   4. liver, adipose tissue, skeletal muscles, neural tissue,
other peripheral tissues;   5. 4;   6. liver;   7. triglycerides;   8. amino, fatty
acids;   9. LDL, HDL;   10. glycerol, fatty acids;   11. glucose, adipose;
12. glycogen;   13. glycogen;   14. gluconeogenesis;   15. acetyl Coa;

16. ketone;   17. acetyl CoA;   18. glucogenic;   19. ketogenic;   20. liver;
21. fatty;   22. ketone;   23. glucose;   24. ketone;   25. Structural;
26. balanced;   27. milk, meat, vegetable and fruit, bread and cereal;
28. complete;   29. milk;   30. N;   31. carbohydrates, lipids;   32. nitrogen;
33. positive;   34. negative;   35. i, b, e, a, g, d, f, j, h, c;   36. k, d, e, m, g, b, i, j, a, c, f, l. h;   37. calorimetry;   38. calorie;   39. 1000;   40. 9.46, 4.32, 4.18;
41. 42, 58;   42. basal;   43. 70;   44. radiation, conduction, convection, evaporation;   45. infrared;   46. half;   47. conduction;   48. convection;
49. heat;   50. perspiration;   51. humidity;   52. biochemical;   53. water;
54. preoptic;   55. heat-loss, heat-gain;   56. a. heat-loss   b. vasodilation
c. radiation   d. convection   e. sweat glands   f. depth;   57. a. decreases
b. radiation   c. convection   d. conduction   e. countercurrent   f. shivering
g. nonshivering   h. thyroxine;   58. behavioral;   59. Acclimatization;
60. brown;   61. surface;   62. fever;   63. pyrogens;   64. Interleukin I;
65. heat-gain;   66. crisis;   67. beneficial;   68. anti-pyretic;   69. c, f, d, a, b,
e   Self test. 1. b;   2. d;   3. c;   4. c;   5. d;   6. c;   7. b;   8. e;
9. e.;   10. d;   11. b;   12. d;   13. b;   14. e;   15. c

Chapter 26. 1. kidneys, ureters, bladder, urethra;   2. respiratory, digestive,
integumentary;   3. kidneys;   4. a. renal artery   b. left kidney   c. urinary
bladder   d. aorta   e. ureter   f. renal vein   g. inferior vena cava;   5. hilus;
6. retroperitoneal;   7. i, k, g, b, e, c, f, j, a, h, d,;   8. arteries;   9. retroperitoneal;
10. 1 liter;   11. rugae;   12. trigone;   13. involuntary;   14. voluntary;
15. nephron;   16. 1.25 million;   17. collecting;   18. a. cortex   b. medulla
c. renal pelvis   d. ureter   e. papilla   f. capsule   g. minor calyx   h. renal column
i. major calyx;   19. tubules;   20. corpuscle, Bowman's;   21. glomerulus;
22. endothelium, lamina densa, capsular;   23. fenestrated;   24. lamina densa;
25. podocytes, pedicels;   26. slit;   27. a. DCT   b. peritubular capillary bed
c. collecting tubule   d. loop of Henle   e. glomerulus   f. renal corpuscle   g. afferent
arteriole   h. efferent arteriole   i. PCT;   28. PCT - loop of Henle - DCT;   29. c,
h, a, d, f, g, e, b;   30. collecting ducts;   31. papillary;   32. afferent, efferent,
peritubular;   33. transitional;   34. transitional, stratified columnar, stratified
squamous;   35. detrusor;   36. a. outer longitudinal muscle   b. circular muscle
c. inner longitudinal muscle   d. trigone   e. uretral openings   f. internal sphincter
g. external sphincter   h. urethra   i. ureters   j. transitional epithelium   k. lumen
l. fibrous coat;   37. filtration, active transport, osmosis;   38. filtrate osmotic
pressure;   39. 10;   40. glomerular;   41. 125, 180;   42. reabsorbed;
43. 80;   44. transport maximum;   45. urine;   46. threshold;   47. electrolytes;
48. peritubular;   49. water;   50. Distal convoluted;   51. actively;
52. sodium;   53. ammonia, ammonium;   54. 1000;   55. osmolarity
(concentration), osmolarity (concentration);   56. isotonic, hypertonic, hypotonic;
57. gradient;   58. collecting;   59. a. impermeable   b. permeable
c. concentrates   d. positive   e. urea   f. water;   60. vasa recta;   61. isotonic;
62. 1200;   63. salts;   64. 100;   65. collecting;   66. hypertonic;
67. ADH;   68. permeable;   69. impermeable;   70. urobilogen;
71. polyuria, oligouria, anuria;   72. azotemia;   73. Clearance;   74. dilate,
constrict;   75. a. increase blood pressure   b. increase glomerular filtration pressure
c. elevation of blood volume;   76. constriction, decrease;   77. peristaltic;
78. micturition;   79. e, c, f, a, b, g, h, d   Self test. 1. e;   2. b;   3. b;   4. e;
5. d;   6. a;   7. b;   8. c;   9. c;   10. e;   11. a;   12. c;   13. d;
14. e;   15. d

Chapter 27. l. ECF, ICF;   2. plasma, interstitial fluid;   3. selectively;
4. sodium chloride, bicarbonate;   5. potassium, magnesium, phosphate;
6. osmolarity;   7. volume, osmolarity;   8. ADH, aldosterone, ANF;   9. a, b, a,
b, c, c, a;   10. 2500;   ll. drinking, eating, metabolism;   12. ICF, ECF;
13. ADH;   14. high;   15. sodium;   16. urine, perspiratiion;   17. potassium,
hydrogen ion;   18. water;   19. increases;   20. ADH;   21. volume;
22. a. volume   b. renin   c. thirst   d. ADH   e. aldosterone   f. sodium g. water;
23. water;   24. cells;   25. secretion;   26. aldosterone;   27. decreased;
28. enzymes;   29. neutral, acid, basic;   30. acids, bases;   31. salts;
32. strong, weak;   33. acid;   34. carbonic;   35. pH;   36. buffer, pulmonary,
renal;   37. buffers;   38. acid;   39. volatile, fixed, organic;   40. carbon
dioxide;   41. carbonic;   42. sulfuric, phosphoric;   43. organic;
44. buffers;   45. weak base;   46. bicarbonate, phosphate, protein;
47. bicarbonate;   48. hydrogen;   49. bicarbonate;   50. a. carbon di
h. superficial dorsal vein;   34. detumescence;   35. LH;   36. testosterone;
37. FSH, inhibin;   38. ovaries, uterine tubes, uterus, vagina;   39. broad
ligament;   40. mesovarium;   41. tunica albuginea, cortex, medulla;
42. infundibulum, fimbriae;   43. cilia;   44. 3- 4;   45. placenta;
46. uterosacral, round, lateral;   47. body, isthmus, cervix;   48. endometrium,
myometrium, serosa;   49. functional;   50. vagina;   51. penis;   52. f, a, d, g,
e, b, c,;   53. areola;   54. lactiferous;   55. a. uterine tube   b. ovary   c. round
ligament   d. pubic symphysis   e. clitoris   f. labium minora   g. labium majora
h. bladder   i. urethra   j. vagina   k. anus   l. cervix   m. uterus   n. fimbria;
56. a. uterine tube   b. round ligament   c. body   d. vagina   e. cervix   f. broad
ligament   g. fimbria   h. ovary   i. ovarian ligament   j. ampulla;   57. a. urethra
b. anus c. vulva d. vagina e. labia minora f. labia majora g. prepuce h. mons pubis
i. clitoris;   58. parasympathetic;   59. monthly;   60. follicles;
61. primordial;   62. primary;   63. zona pelucida;   64. microvilli;
65. secondary;   66. tertiary;   67. corona radiata;   68. corpus luteum;
69. corpus albicans;   70. ovarian;   71. FSH;   72. estrogen;   73. FSH, LH;
74. LH;   75. corpus luteum;   76. progesterone;   77. GnRH;   78. FSH;
79. m, k, i, j, e, f, d, d, c, c, a, a, l, b, g, h, n,;   80. 50;   81. primordial;
82. estrogen, progesterone;   83. 8, l, 7, 6, 3, 2, 5, 4,;   84. g, d, a, b, e, f, c, h   Self
test l. d;   2. b;   3. b;   4. d;   5. d;   6. b;   7. c;   8. a;   9. d;
10. c;   ll. d;   12. b;   13. b;   14. e;   15. c

Chapter 29. l. development;   2. differentiation;   3. fertilization;
4. embryogenesis;   5. fetal;   6. postnatal;   7. differentiation;   8. inheritance,
genetics;   9. b, c, b, d, a, f, g, e;   10. 30;   ll. primary spermatocytes;
12. reductional;   13. equational;   14. diploid, haploid;   15. primary oocytes;
16. polar body;   17. polar body;   18. four;   19. three;   20. a. primary oocyte
b. first polar body   c. second and third polar bodies   d. ovum   e. secondary oocyte
f. spermatozoa   g. spermatids   h. secondary spermatocytes   i. primary
spermatocyte;   21. nutrition;   22. uterine tube;   23. hyaluronidase;
24. activated;   25. amphimixis;   26. gentotype, phenotype;   27. homologous;
28. alleles;   29. homozygous, heterozygous;   30. dominant;   31. recessive;
32. A, a;   33. Punnet's;   34. autosomes, sex;   35. XY, XX;   36. X, X, Y;
37. autosomal;   38. sons - 50% XCY, normal, 50% XcY color blind daughters 50%
XCXC normal, 50% XCXc carriers (but normal vision);   39. induction;
40. genes;   41. cleavage, implantation, placentation, embryogenesis;
42. cleavage;   43. blastomere, morula;   44. blastocoel;   45. trophoblast,

inner;   **46.** implantation;   **47.** gastrulation;   **48.** extraembryonic;   **49.** yolk sac, allantois, chorion, amnion;   **50.** endoderm, mesoderm;   **51.** blood; **52.** ectoderm, mesoderm;   **53.** fluid;   **54.** urinary bladder;   **55.** chorion; **56.** chorionic;   **57.** decidua basalis;   **58.** 90;   **59.** organs;   **60.** third trimester;   **61.** a. chorion   b. chorionic villus   c. allantois   d. yolk sac e. umbilical cord   f. embryo   g. amniotic cavity   h. amnion;   **62.** 6, 5, 1, 7, 8, 1, 4, 8, 5,;   **63.** 50, 50;   **64.** parturition;   **65.** labor;   **66.** dilation, expulsion, placental;   **67.** cervix;   **68.** amnion;   **69.** expulsion;   **70.** placental; **71.** breech;   **72.** dizygous (fraternal);   **73.** monozygous (identical); **74.** cojoined;   **75.** neonatal, infancy, childhood, adolescence, maturity; **76.** neonatal;   **77.** infancy, childhood;   **78.** growth;   **79.** a. decline in populations of cells.   b. decreased ability to replace dividing cells.   c. genetic changes.   d. mutations;   **80.** accidents;   **81.** c, b, f, a, d, g, e,   Self test. **1.** d; **2.** d;   **3.** b;   **4.** c;   **5.** b;   **6.** d;   **7.** c;   **8.** a;   **9.** c;   **10.** a;   **11.** c; **12.** b;   **13.** e;   **14.** b;   **15.** d